Please send any correspondence regarding permissions to:
UniversityOfHealthCare
419 N. Larchmont Blvd., #323
Los Angeles, CA 90004



ISBN: 159491057X

Library of Congress Catalog Number: 2004112052

UniversityOfHealthCare website: www.uohc.com

Neither the author nor the publisher assumes any liability for the information contained in this title. Every reader must consult the latest professional information available at the time and employ it in consultation with a professional who is familiar with the specific situation.

Contents

# Introduction

This product consists of a combination of this manual and a CD. The manual is a summary of the important points presented more interactively on the CD.

When you place the CD in your computer, it should load automatically and install with some simple steps. If it does not start automatically, open up Windows Explorer (from Start, Programs, Accessories) and navigate to the setup file with the abstract arrow icon on your CD drive. Then double click it. It will then install on your computer. After that, you may open the program by clicking on Start, Programs, and then the name of the program. This will open the program file. After you run it the first time, an icon will automatically be added to your desktop.

This CD program is protected with anti-piracy software. The library edition is licensed to one user on one computer for one time for borrowers from a library. As a result, if you decide to purchase the product, you will need to follow the instructions after you install the CD. One complimentary activation is allowed per library for a computer at the library. The request must come from an email with the library URL in the return address.

You must have Internet Explorer 4.0 or higher running on your computer. Supported operating systems are Windows 95, 98, 98 SE, ME, 2000, or XP.

For any questions about your license or the program, contact us at license@uohc.com.

About the Authors

M. Daniel Farb, CEO of UniversityOfHealthCare and UniversityOfBusiness, is a leader in the field of interactive management and healthcare e-learning. He received a BA in English Literature from Yale (where he set an academic record and studied with writers like Robert Penn Warren), an M.D. from Boston University, a degree in Executive Management from the Anderson School of Business at UCLA, and is currently working on a degree at UCLA in International Trade. He is a practicing ophthalmologist. He also has received two patents in ophthalmology and is working on others, has worked with the World Health Organization in Geneva and the National Institutes of Health in Washington, D.C. He has written scientific and popular articles, and has worked as a newspaper reporter. He helped Dr. Robbins edit one of the editions of Robbins' "Pathology" textbook for readability. He wrote an article on humor for the Massachusetts Review. He has experience in theater and television, including acting, directing, and stage-managing. He has programmed his own patient records database. He has written and edited hundreds of e-learning courses.

Dr. Farb is a member of the American Academy of Ophthalmology, the Union of American Physicians and Dentists, the AOJS, the American Association of Physicians and Surgeons, the ASTD (American Society for Training and Development), the E-Learning Forum, the Southern California Biomedical Council, the PDA (Parenteral Drug Association), and the Medical Marketing Association.

BRUCE GORDON is the Creative Director for UniversityOfHealthCare, LLC, and plays an important role in writing the more creative projects, especially those with stories.

After receiving a BA in Economics from UCLA, he began a freelance writing career that included technical writing (such as a manual for Princess Cruise Lines), stand-up comedy routines for nationally known comedians, and screenplay writing.
He has done production support work with famous Hollywood personalities on such well-known productions as Aaron Spelling's "Dynasty" and "Love Boat" TV shows.

An audio-visual software specialist, he is a versatile artist, with published works in a variety of media, including music, motion graphics, and digital video short film.

# OSHA Occupational Radiation Safety

Just what is "radiation"?
Energy coming from some source that travels through any material or through space. Examples of radiation are: Light, heat, and sound.

Ionizing radiation

Ionizing radiation is another kind of radiation, which involves the production of charged particles (ions) in matter. We will focus on this type of radiation.

Some forms of ionizing radiation results from cosmic rays and from naturally occurring radioactive materials existing in the earth and in live organisms.

Colorless, odorless, and hence undetectable by the senses, radiological materials must be recognized with the help of specifically-designed equipment.

A variety of determinations need to be made:

- Whether or not the level of radioactivity presents an immediate health hazard.
- The size of the affected area
- Whether or not the level of radioactivity presents a long-term health hazard.

Because contaminated people and animals sometimes migrate before the symptoms of the radiological incident are discovered, the affected area could be greater than that which is readily apparent.

Radiation exposure is measured in "rem" units and "millisieverts," where 10 rem = (100 mSv).

Occupational radiation exposure is governed by OSHA (the Occupational Safety and Health Administration.

One of their main publications of regulations (relevant to this course) is entitled, "Occupational Safety and Health Standards-Toxic and Hazardous Substances- Ionizing Radiation, Regulations 29 CFR, Standard No. 1910-1096".

What is occupational radiation exposure?
Occupational radiation exposure is exposure that happens to workers during the workday and during the course of doing their jobs.

What industries subject workers to occupational exposure with ionizing radiation?
-Mining and milling
-Fuel processing and fabrication facilities
-Educational and research establishments
-Nuclear fuel cycle facilities
-Medical institutions

What does OSHA consider radioactive material?
All substances that emit corpuscular (discrete particles) or electromagnetic emanations (emissions) by spontaneous nuclear disintegration (the natural or induced transformation of an atomic nucleus from a more massive to a less massive configuration by the emission of particles or radiation).

What does OSHA consider a restricted area?
Any area to which access is regulated by the employer so as to protect persons from exposure to radioactive substances or radiation.

What does OSHA mean by dose?
The amount of ionizing radiation absorbed, per unit of mass, by the body or by any portion of the body during a period of time. Several different units of dose are in current use.

What units does OSHA use to measure ionizing radiation dosage?
-Rad
-Rem

What does Rad mean?
It's a measure of the dose of any ionizing radiation to body tissues in terms of the energy absorbed per unit of mass of the tissue. One millirad (mrad)=0.001 rad.

What does Rem mean?
It's a measure of the dose of any ionizing radiation to body tissue in terms of its estimated biological effect relative to a dose of 1 roentgen (r) of X-rays (1 millirem (mrem)=0.001 rem).

What does air dose mean?
It means that a properly calibrated appropriate instrument in air measures the dose. This measurement is taken at or near the body surface in the region of the highest dosage rate.

What are OSHA's regulations for radiation exposure in restricted areas?
An employer is not allowed to possess, use, or transfer sources of ionizing radiation in any way that could make any person in a restricted area receive dangerous doses:
-In any period of one calendar quarter
-From sources in the employer's possession or control
-In excess of the established safe limits

What are OSHA's established safe limits (measured in rems per calendar quarter)?

- 1 ¼ rems per calendar quarter for the whole body (includes the head and trunk, active blood-forming organs, lens of eyes, or gonads)
- 18 ¾ rems per calendar quarter for hands and forearms. Same for feet and ankles
- 7 ½ rems per calendar quarter for skin of the whole body

What conditions must be met for an employer to allow a person in a restricted area to receive whole body doses greater than the established safe limits just detailed?
-If the dose to the whole body doesn't exceed 3 rems during any calendar quarter.
-If the dose to the whole body, when added to the accumulated occupational dose to the whole body, doesn't exceed 5 (N-18) rems (where "N" equals the individual's age in years at his last or her birthday).
-If the employer maintains adequate past and current exposure records that show that the addition of such a dose will not cause the individual to exceed the authorized amount.

What is the regulation for employees under the age of 18?
An employer is not allowed to permit any employee who is under 18 years of age to receive more than 10 percent of the adult limits specified above (in any period of one calendar quarter).

What are the regulations for exposure to airborne radioactive material?
An employer is not allowed to possess, use or transport radioactive material in any way that would cause an employee, within a restricted area, to be exposed to airborne radioactive

material in an average concentration exceeding the limits specified in Table 1 of appendix B to 10 CFR part 20.

What period is considered when limiting airborne radioactive material exposure?
-40 hours in any work week of 7 consecutive days.
-If the number of exposure hours is less than 40, then the limits specified in the table are allowed to be increased proportionately.
-But if the number of exposure hours is greater than 40, the limits specified in the table have to be <u>decreased</u> proportionately.

What period is considered for persons under 18 when limiting airborne radioactive material exposure?
-One week
-The limits are specified in Table II of appendix B to 10 CFR part 20.
-Concentrations may be averaged over periods not greater than 1 week.

What does OSHA specify as precautionary procedures and personal monitoring?
Every employer has to make as many radiation surveys as are necessary so as to comply with the regulations' provisions.

What is a survey?
An evaluation of the radiation hazards incident to the production, use, release, disposal, or presence of radioactive materials. This also applies to other sources of radiation under a specific set of conditions.

When it's appropriate, this type of evaluation includes:

-A physical survey of the location of materials and equipment
-Measurements of levels of radiation or concentrations of radioactive material present.

What personnel monitoring equipment do employers have to provide workers with?
-Film badges
-Pocket chambers
-Pocket dosimeters
-Film rings

Which employees are required to use personnel monitoring equipment?
-Every employee who enters a restricted area that would likely cause that worker to receive a radiation dose in any calendar quarter in excess of 25 percent of the safe limits.
-Every employee under 18 years of age who enters a restricted area that would likely cause that worker to receive a radiation dose in any calendar quarter in excess of 5 percent of the safe limits.
-Every employee who enters a high radiation area.

What, specifically, does OSHA mean by the term radiation area?
Any worker-accessible area that has radiation at such levels that a major portion of the body could receive more than a 5 millirem dose in any hour, or a 100 millirem dose in any 5 consecutive days.

What does OSHA mean when it speaks of a high radiation area?
Any worker-accessible area that has radiation at such levels that a major portion of the body could receive more than 100 millirem in any one hour.

## CAUTION SIGNS

What is a general OSHA rule for symbols on radiation caution signs, labels, and signals?
Symbols have to use the conventional radiation caution colors (magenta or purple on yellow background). The symbol is the conventional three-bladed design:
Radiation Symbol
1) Cross-hatched area to be magenta or purple.
2) Background is to be yellow.

How should radiation caution signs be worded and located in a radiation area?
Each radiation area needs to be conspicuously posted with a sign or signs bearing the radiation caution symbol and the words:

CAUTION

RADIATION AREA

How should radiation caution signs be worded and located in a high radiation area?
Each high radiation area needs to be conspicuously posted with a sign or signs bearing the radiation caution symbol and the words:

CAUTION

HIGH RADIATION AREA

What additional provisions must be made for each high radiation area?
-Every high radiation area needs to be equipped with a control device that can reduce radiation below the level at which a

worker might receive a dose of 100 millirems in 1 hour upon entry into the area.

- Or, every high radiation area needs to be equipped with a device that can energize a conspicuous visible or audible alarm signal in a way that causes the entering worker and the employer or a supervisor of the activity to be made aware of the entry.

-This type of control device isn't required if the high radiation area has only been established for 30 days or less.

What are the OSHA signage rules for airborne radioactivity areas?

Every airborne radioactivity area needs to be conspicuously posted with a sign or signs bearing the radiation caution symbol and the words:

CAUTION

AIRBORNE RADIOACTIVITY AREA

Every area or room where radioactive material is used or stored containing radioactive material (other than natural uranium or thorium) in amounts greater than 10 times the safe limits (listed in appendix C to 10 CFR part 20) needs to be conspicuously marked with a sign or signs bearing the radiation caution symbol and the words:

CAUTION

RADIOACTIVE MATERIALS

Every area or room where natural uranium or thorium is used or stored in amounts greater than 100 times the safe limits (listed in 10 CFR part 20) needs to be conspicuously marked with a sign or signs bearing the radiation caution symbol and the words:

CAUTION

RADIOACTIVE MATERIALS

Each container employed to transport, store, or use radioactive material (other than natural uranium or thorium) in amounts greater than the safe limits (listed in appendix C to 10 CFR part 20) needs to be affixed with a durable, clearly visible label bearing the radiation caution symbol and the words:

CAUTION

RADIOACTIVE MATERIALS

Each container used to transport, store, or use natural uranium or thorium in amounts greater than 10 times the safe limits (listed in appendix C to 10 CFR part 20) needs to be affixed with a durable, clearly visible label bearing the radiation caution symbol and the words:

CAUTION

RADIOACTIVE MATERIALS

When is a label not required?

-If the concentration of the material in the container isn't greater than the safe limits (specified in column 2 of Table 1 of appendix B to 10 CFR part 2).

-For laboratory containers, such as beakers, flasks, and test tubes, used transiently in laboratory procedures (when the user is present).

Where containers are used for storage, what else must the labels state?

-The quantities and kinds of radioactive materials in the containers
-The date of measurement of the quantities.

What are the proper signal characteristics of an immediate evacuation warning signal?
-The signal needs to be a midfrequency complex sound wave amplitude modulated at a subsonic frequency. The complex sound wave in free space needs to have a fundamental frequency (f(1)) between 450 and 500 hertz (Hz) modulated at a subsonic rate between 4 and 5 hertz.
-The signal generator can't be less than 75 decibels at every location where any person might be present whose immediate, rapid, and complete evacuation is essential.
-A sufficient number of signal units need to be installed where an individual might be present whose immediate, rapid, and complete evacuation is essential.
-The signal needs to be unique in the plant or facility in which it is installed.
-The minimum duration of the signal needs to be sufficiently long enough to insure that all affected persons hear the signal.
-The signal-generating system needs to respond automatically to an initiating event without requiring any human action to sound the signal.
-The signal-generating system needs to be designed to incorporate components that enable the system to produce the desired signal each time it's activated within one-half second of activation.
- Just in case operation during power failure is necessary, the signal-generating system needs to be provided with an automatically activated secondary power supply that's adequate to power simultaneously all emergency equipment to which it is connected. Systems using batteries as the primary source of power are exempt from this requirement.

-All components of the signal-generating system need to be located to provide maximum practicable protection against damage in case of fire, explosion, corrosive atmosphere, or other environmental extremes consistent with adequate system performance.
-The signal-generating system needs to be designed with the minimum number of components necessary to make it function as intended, and it should use components that don't require frequent servicing (such as lubrication or cleaning).
-Where several activating devices feed activating information to a central signal generator, failure of any activating device can't render the signal-generator system inoperable to activating information from the remaining devices.
-The signal-generating system needs to be designed in a way that enhances the probability that alarm occurs only when immediate evacuation is warranted. The number of false alarms can't be so great that the signal will come to be disregarded and they have to be low enough to minimize personal injuries or excessive property damage that might result from such evacuation.
-Initial tests, inspections, and checks of the signal-generating system have to be made:
A) To verify that the fabrication and installation were made in accordance with design plans and specifications
B) To develop a thorough knowledge of the performance of the system and all components under normal and hostile conditions.

-Once the system has been placed in service, periodic tests, inspections, and checks need to be made to minimize the possibility of malfunction.
-Following significant alterations or revisions to the system, tests and checks similar to the initial installation tests have to be made.

-Tests need to be designed to minimize hazards while conducting the tests
-Prior to normal operation, the signal-generating system needs to be checked physically and functionally to assure reliability and to demonstrate accuracy and performance.

What needs to be included in specific tests?
-All power sources
-Calibration and calibration stability
-Trip levels and stability
-Continuity of function with loss and return of required services such as AC or DC power, air pressure, etc.
-All indicators
-Trouble indicator circuits and signals, where used
-Air pressure (if used)
-A determination that the signal sound level within the required limits at all points that require immediate evacuation

What periodic scheduled performance tests and status checks are required in addition to the initial startup and operating tests?
Periodic scheduled performance tests and status checks must be made to insure that the system is at all times operating within design limits and capable of the required response.

What's one thing the periodic tests or checks should include?
The adequacy of the signal activation device. This includes:
-All power sources.
-Function of all alarm circuits and trouble indicator circuits including trip levels.
-Air pressure (if used).
-Function of entire system including operation without power where required.

What's another thing that should be included in the periodic tests?
Complete operational tests including sounding of the signal and determination that sound levels are adequate.

How often should periodic tests be conducted?
Periodic tests should be scheduled on the basis of need, experience, difficulty, and disruption of operations. The entire system needs to be operationally tested at least quarterly.

What training should workers be given in regard to the signal?
All employees whose work may necessitate their presence in an area covered by the signal need to be made familiar with the actual sound of the signal—preferably as it sounds at their work location.

Before placing the system into operation, all employees normally working in the area need to be made acquainted with the signal by actual demonstration at their work locations.

What are the exemptions from posting requirements?
- If the radiation level 12 inches from the surface of a sealed-source container or housing doesn't exceed 5 millirem per hour, a room or area doesn't have to be posted with a caution sign.
- If personnel are in attendance who will take the precautions necessary to prevent people from becoming exposed to radiation or radioactive material in excess of the established safe limits, rooms or other areas in onsite medical facilities aren't required to be posted with caution signs.

What is another specific condition for exemption from the posting requirement?

Caution signs aren't required to be posted at areas or rooms containing radioactive materials for periods of less than 8 hours provided that:
-The materials are constantly attended during such periods by someone who'll take the precautions necessary to prevent anyone from being exposed to radiation or radioactive materials in excess of the safe limits
-And such area or room is subject to the employer's control.

What are the exemptions for radioactive materials packaged for shipment?
Radioactive materials packaged and labeled in accordance with regulations of the Department of Transportation (published in 49 CFR Chapter I), are exempt from the labeling and posting requirements of this subpart during shipment, as long as the inside containers are labeled accordingly.

What regulations do employers have to follow for instructing personnel and for posting of occupational radiation information?
-Employers regulated by the Nuclear Regulatory Commission are governed by 10 CFR part 20 standards.
-Employers in a State shall be governed by the requirements of the laws and regulations of that State.

If an employer isn't regulated by either the Nuclear Regulatory Commission or a State, then what are the requirements for instructing personnel and for posting of occupational radiation information?
All individuals working in or frequenting any portion of a radiation area have to be:
-Informed of the occurrence of radioactive materials or of radiation in such portions of the radiation area

-Instructed in the safety problems associated with exposure to such materials or radiation and in precautions or devices to minimize exposure
-Instructed in the applicable provisions of this section for the protection of employees from exposure to radiation or radioactive materials
-Advised of reports of radiation exposure which employees may request pursuant to the regulations in this section.

What other requirements exist for instructing personnel and posting of occupational radiation for employers not regulated by either the Nuclear Regulatory Commission or a State?
-Each such employer has to post a current copy of its provisions and a copy of the operating procedures applicable to the work.
-The employer's postings have to be easy to see and located where they will insure all employees working in or frequenting radiation areas will see the documents as they come to and from the workplace.
-The employer needs to keep these documents available for workers to examine them upon request.

Radioactive materials stored in a non-radiation area have to be secured against unauthorized removal from the place of storage. No employer is permitted to dispose of radioactive material except by transfer to an authorized recipient, or in a manner approved by the Nuclear Regulatory Commission or a State.

When is an immediate notification of incident required for employees not protected by the Nuclear Regulatory Commission or the requirements of the laws and regulations of States?
Such an employer must immediately notify the Assistant Secretary of Labor or his duly authorized representative (by telephone or telegraph) of any incident involving radiation

which may have caused or threatens to cause one or more of the following:
-Exposure of the whole body of any individual to 25 rems or more of radiation
-Exposure of the skin of the whole body of any individual to 150 rems or more of radiation
-Exposure of the feet, ankles, hands, or forearms of any individual to 375 rems or more of radiation
-The release of radioactive material in concentrations which, if averaged over a period of 24 hours, would exceed 5,000 times the limit specified for such materials in Table II of appendix B to 10 CFR part 20.

Within 24 hours following its occurrence, the employer has to notify the Assistant Secretary of Labor or his duly authorized representative (by telephone or telegraph) of any incident involving radiation which may have caused or threatens to cause any of the following:
-Exposure of the whole body of any individual to 5 rems or more of radiation
-Exposure of the skin of the whole body of any individual to 30 rems or more of radiation
-Exposure of the feet, ankles, hands, or forearms to 75 rems or more of radiation

## OVEREXPOSURE

To whom (and when) do employers need to make a written report concerning overexposure and excessive levels and concentrations?
Employers need to make a written report within 30 days to the Assistant Secretary of Labor or an authorized representative any and every time an individual is exposed to radiation or radioactive material in excess of the safe limits.

What specific information needs to be described in the written report of worker radiation exposure?
-The extent of exposure of persons to radiation or to radioactive material
-The levels of radiation and concentration of radioactive material involved
-The cause of the exposure, levels of concentrations
-The corrective steps taken or planned to assure against a recurrence
-If a case requires that an employer should report worker radiation exposure to the U.S. Department of Labor, then the individual must also be notified in writing of the nature and extent of exposure.
-This notice has to contain the following statement: "You should preserve this report for future reference."

* What is every employer required to do when it comes to worker's exposure records?
-Maintain records of the radiation exposure of all employees for whom personnel monitoring is required
-Advise each worker of his or her individual exposure at least annually.
-Maintain records in the same units used in this section of the regulations.
-If a former employee requests a written report of his or her radiation exposure as shown in an employer's records, that employer must furnish those records to the former worker.
-The report has to be furnished within 30 days from time of the request, and it has to cover each calendar quarter of the individual's employment involving exposure to radiation. If the employee requests a shorter time period for the report, the employer must comply.

-The report has to include the results of any calculations and analysis of radioactive material deposited in the employee's body.
-The report has to be in writing, and it has to include the following statement: "You should preserve this report for future reference."

Why should special considerations be given to pregnant workers in regards to occupational radiation?
- Because an embryo or fetus is especially sensitive to radiation since the cells and tissues of an embryo or fetus are rapidly growing and dividing.
-The embryo/fetus is at the most radiosensitive stage of development, especially during the first 20 weeks of pregnancy.

What is the current threshold dose for developmental effects? About 10 rem, but throughout the whole gestation period, the fetus shouldn't get more than 500 mrem.

What should a radiation worker do if she becomes pregnant?
-Take extra precautions to ensure the least possible radiation exposure to the embryo/fetus.
-Notify the supervisor as soon as a pregnancy is suspected or known.
-Wear additional shielding, such as a lead apron
-Wear a secondary dosimeter badge around the waist and check it more often (i.e. monthly rather than quarterly).

## RADIATION FUNDAMENTALS

Unstable atoms produce ionizing radiation.
How are unstable atoms different from stable atoms?
Unstable atoms have an excess of energy or mass or both.

That's what makes them radioactive, since the emissions of their excess energy or mass as they try to reach stability is called radiation.

These are the different kinds of ionizing radiation:

- Electromagnetic (like light). Radiation examples: Gamma radiation & x-rays
- Particulate (mass given off with the energy of motion). Radiation examples: Beta and alpha radiation.

Devices such as x-ray machines can also produce ionizing radiation. When medical personnel use machines like an x-ray machine, ionizing radiation is produced. The radiation exposure only happens when the device is turned on.

The following are facts about gamma radiation and x-rays:
A) Like visible light, radio waves, and ultraviolet light, gamma radiation and x-rays are forms of electromagnetic radiation. But gamma rays and x-rays have much more energy than the other types of electromagnetic radiation.
B) Because gamma radiation easily penetrates many substances, it is sometimes referred to as "penetrating radiation". It can travel many meters in air and many centimeters in human tissue.
C) X-rays are also penetrating radiation.
D) Radioactive substances that give off gamma radiation and x-rays present a health hazard to humans, both externally and internally.
E) Gamma radiation can only be shielded with dense materials. Penetrating radiation can't be shielded by clothing or and turnout gear, although but those things can prevent contamination of the skin by radioactive materials.
F) Civil defense and other survey instruments can detect gamma radiation. A device called a Geiger counter (e.g. CD V-700, detailed later) can measure low levels. When the levels

are high, they're measured with an instrument called an ionization chamber (e.g. CD V-715).
G) When a radioactive material gives off alpha and beta radiation, gamma radiation or x-rays often accompany those emissions.
H) Gamma radiation can't be detected by devices made exclusively for alpha detection, like an alpha scintillation counter.
I) Accumulated exposure to gamma radiation is measured with instruments known as dosimeters. These can include film badges, pocket chamber (pencil) and thermoluminescent dosimeters.

Beta Radiation
A) Beta radiation is somewhat penetrating and can travel meters in air.
B) Beta radiation is able to penetrate to the layer of human skin in which new skin cells are generated (the germinal layer). Contaminants that give off beta radiation can cause skin injury if they stay in contact with skin for an extended amount of time.
C) Internally deposited beta radiation contaminants can cause injury.
D) Survey instruments like a CD V-700 can recognize beta-emitting contaminants, if the metal probe cover is open). But there are a few beta emitters, such as sulfur-35, carbon-14, and tritium that are hard to detect. These substances generate very low energy and radiation that penetrates poorly.
E) Ionization chambers (e.g. a CD V-715) can't recognize beta radiation.
F) Some protection from beta radiation is afforded by clothing and turnout gear. Beta emitters can be kept off the skin with turnout gear and dry clothing.

Alpha Radiation

A) Alpha radiation can't penetrate skin.
B) It's dangerous for humans to inhale, swallow, or absorb alpha-emitting materials through open wounds.
C) To measure alpha radiation, a number of instruments can be used. However, it's imperative for accurate measuring that users receive special training.
D) Alpha radiation-emitting substances can't be detected by civil defense instruments (e.g. CD V-700), unless beta and/or gamma radiation are also being emitted.
E) Since alpha radiation isn't penetrating, it can't be detected by instruments through even a thin layer of materials such as paper, dust, blood, or water.
F) Alpha radiation can only travel through air for an extremely short distance.
G) Clothing, turnout gear, and covers on probes can't be penetrated by alpha radiation. Alpha emitters can be kept off the skin with turnout gear and dry clothing.

There is a radioactive chemical that is used extensively for commercial and research functions, known as "Strontium-90." Strontium is actually a radioactive chemical element that's found naturally in the environment.

What are some of the uses of Strontium 90?
* As a radioactive tracer in medical and agricultural instruments
* In bone cancer treatment
* To produce glass for color television tubes.
* To make electricity from the heat it generates, for weather stations, satellites, and navigational beacons.

What are some of the characteristics of Strontium-90?
* It's a soft, silvery metal
* When it reacts with air, it turns yellow.

* In nature, 4 stable isotopes of strontium are found (only the number of neutrons are different among isotopes).
* A by-product of uranium and plutonium fission in nuclear reactors,

Strontium-90 is the most important radioactive isotope.

* Strontium-90 has a half-life of 29 years ("half-life" means how much time is necessary for 1/2 of the isotope to decay).

How have people been exposed to Strontium-90, even without resulting from a bioterrorism dirty bomb?
Massive quantities were released during the nuclear testing of the 1950's and 1960's. It appears that today's levels from those tests are low. The Soviet nuclear accident at Chernobyl caused atmospheric dispersal of Strontium-90, not only throughout nations in the Soviet Republics and Northern Europe, but also lesser amounts worldwide.

The Chernobyl disaster will be detailed later in the "Past Catastrophes" section.

Residents within close proximity to nuclear power plants get exposed to a greater level of Strontium-90 than people who don't live near reactors. Because Strontium-90 has been dispersed into the environment, everybody has some level of exposure through food, water, and by inhalation.

How can Strontium-90 exposure damage humans?
It can be readily absorbed by human bones and blood-forming tissues, since its structure is similar to that of calcium. This can result in bone cancer, cancer of the soft tissue surrounding the bone, and leukemia.

The more Strontium-90 absorbed, the greater the chances of cancer. Other conditions include a person's physical conditions while being exposed, and Strontium-90's concentration in the environment. Children from birth to one year old typically absorb Strontium-90 at a rate about 60 percent greater than adults, but the absorption rate for adults can increase to similar levels with low-calcium diets and fasting.

Some industries intrinsically have more exposure to Strontium-90 (and in higher amounts) than most others.
* Scrap metal sorting, sales and brokerage
* Metal melting and casting
* Municipal landfill operations

Regardless of where or how an accident involving radiation happens, three types of radiation-induced injury can occur:

1. External irradiation
2. Contamination with radioactive materials
3. Incorporation of radioactive material into body cells, tissues, or organs

It's possible that all three kinds of exposure can occur simultaneously. Physical injury or illness can add to the problem of radiation exposure.

Whenever presented with serious medical problems, radiation concerns such as radiation monitoring, contamination control, and decontamination become secondary concerns.

When does external irradiation occur?
When part of the body (or the whole thing) is exposed to penetrating radiation from an external, unshielded source. The radiation during this exposure can either pass all the way through the body or be absorbed by it.
Example: This happens during a normal chest x-ray.

External irradiation does not make a person radioactive, and following external exposure, an individual can be treated like any other patient.

The second kind of radiation damage results when a victim is contaminated with radioactive substances.
People get contaminated externally, internally, or in both ways when radioactive substances are released into their environment as gases, liquids or solids.
Radioactive substances can contaminate the skin and other external body surfaces. They can be come deposited internally if they enter the body.
If material containing radioactive atoms is deposited anywhere it isn't wanted (including clothing or skin), this is considered to be contamination.
It is radioactive contamination that can spread.
But radiation itself CANNOT spread; nor does it get "on" or "in" humans.
* If a radioactive substance gets on a person's skin or clothes, then that person is externally contaminated.
* If a person takes in a radioactive substance by breathing it in, swallowing it, or absorbing it through wounds, that person is internally contaminated.
* If the radioactive material is scattered around or uncontained, then it is the environment that is contaminated.
What kind of contamination happens when a radioactive substance gets on a person's skin or clothes?
External Contamination

(But if a person takes in a radioactive substance by breathing it in, swallowing it, or absorbing it through wounds, that person is internally contaminated.)

Now we move on to the 3rd type of radiation injury: incorporation of radioactive material. Just what is incorporation?
Incorporation is the uptake of radioactive materials by body cells, tissues, and target organs, for example: bone, liver, thyroid, or kidney.

A Geiger counter is the most common instrument used for detecting radiation.

How do you begin a survey with a typical Geiger-Mueller (GM) counter?
Remove the GM counter from storage and get fresh batteries.
Set up the device.
Establish what the background level is.
Angle the Geiger counter so that the meter is away from you.
When you find the battery compartment, open it.
Next, insert the batteries in the meter using the correct up or down orientation. Then close and latch the battery compartment.
Check the batteries using the "range" switch or "bat" button; the method depends on the type of instrument.
The meter needle should move to area on scale marked battery, indicating the batteries are good.
If the batteries are not good, find a flashlight or other source of 2 D-cells and put them in the meter -- check these batteries also.
Turn the "F/S" switch to "S" (Slow) and turn the "audio" switch to "ON."

How do you measure the background radiation?
Make sure that the "F/S" switch is on "S" (Slow).
Next, position the range switch on the most sensitive level.
If a probe cover is in place, remove it.

Make a written notation of the reading as you measure the background radiation for 60 seconds. You probably should make several counts and then average the results, because background radiation changes over time.
Once again, make a written record of the reading.

You should anticipate a reading of 40-100 counts/min or a reading of approximately 0.02 mR/hr (i.e. 0.2 on the 0.1 range setting), or 0.2 micro Sv/hr. As always, make a written record of the background reading.

This is how to survey using the instrument:
Change the "F/S" switch to "F" (Fast response). Then position the instrument selector switch to the most sensitive range of the device.
As you maintain a distance of about 1/2 to 1 inch between the probe and the person's skin, survey the whole body from the scalp to the feet without missing any areas.
Then slowly shift the probe's position (approx. 1 inch per second).
Make sure that the probe doesn't touch anything, and strive to keep a constant distance. It's important to give extra attention to the face, hands, and feet.
-- Pay particular attention to hands, face and feet.
It's also important to keep in mind that some GM instruments aren't able to recognize alpha radiation and some low-energy beta radiation. Alpha radiation can't be detected through even a thin film of blood, dirt, water, clothing, or through a probe cover.
An increase in count rate or exposure rate above background indicates the presence of radiation.
Locate the point that results in the most clicks. Switch the "F/S" switch to "S" to take a reading at this location. Keep in

mind that you need to reset it to "F" before you proceed with the survey.
Move the range selector switch to adjust the device's range, if you need to.
Make a written record of the time and radiation instrument readings.

Surfaces that register more than 2 times the previously recorded background level are ordinarily considered contaminated. When the spread of radiation involves alpha emitters, a reading that is less than 2 times the background radiation means that the victim's contamination is not meaningful (from a medical perspective).

Always consult with a health physicist when the situation is suspected of involving a contaminant with an alpha emitter (such as plutonium) or low energy beta emitter.

How do you end the radiation survey?
Turn the meter off.
Put the cap back on the meter probe.
Remove the batteries.
Replace the Geiger counter in its storage case.

What is the process for monitoring personnel?
A) Stand the individuals on a clean pad.
B) Ask the individual to stand up straight with feet slightly spread. Have them keep their arms extended up with their fingers straight out.
C) Make sure that both hands and arms are monitored, then have them turn over their hands so you can repeat the process.
D) Cover the whole body by beginning at the top of the head, then going down from there. Make sure to carefully monitor forehead, nose, mouth, neckline, torso, knees, and ankles.

E) Turn the person around so that you can repeat the survey on back side of their body.
F) Finish by monitoring the soles of the individual's feet.

Whenever a radiation incident occurs, it's best to know what type of radioactive materials were involved. Packaging labels help to cut down on the guess work in making such a determination. Shipments of radioactive materials carry two identifying warning labels secured to opposite sides of the outer package.

Excluded from this labeling system are shipments that only have limited amounts of radioactive substances, or those with low specific activity (LSA).

The following are examples of the three different labels affixed to the external surface of packages containing radioactive substances.

Almost no radiation--0.5 mrem/hr (5 μSv/hr) maximum on surface

Low radiation levels--50 mrem/hr (0.5 mSv/hr) maximum on surface; 1 mrem/hr (10 µSv/hr) maximum at 1 meter

Higher radiation levels--200mrem/hr (2 mSv/hr) maximum on surface; 10 mrem/hr (0.1 mSv/hr) maximum at 1 meter
Also required for fissile class III or large-quantity shipments, regardless of radiation level.

Labels of radioactive substances carry the U.N. (United Nations) hazard class "7".

They specify both the radioactive content and the quantity in measurement units called "curies."

Typical radioactive material warning placard: Standard size is 10 x 10 inches. The placard shown must be used anytime a vehicle carries one or more packages of a Radioactive Yellow III label or if the vehicle is operating under exclusive use provisions required for certain LSA shipments or packages with higher than normal radiation levels.

Any four-digit ID number shown on an adjacent **orange panel** is used for specific identification of the cargo. Package labels and documents, such as shipping papers or bills of lading, which are of great value in assessing potential hazards in transportation accidents or intentional incidents, accompany most shipments of radioactive material. These papers will have a 24-hour contact number for information about the material and potential health hazards.

Yellow-II and Yellow-III labels also specify the transport index.

The number given indicates the maximum radiation level (in mrem/hr) at a distance of one meter from the external surface of a package or container. (Readings in mSv/hr are multiplied by 100 to get mrem/hr.) For example, a TI of 3 (as shown below) would indicate that, at one meter from the labeled package, the radiation intensity that can be measured is no more than 3 mrem/hr (.03 mSv/hr). If the radiation level at one meter from a package is found to be higher than the specified value, a radiation authority should be consulted. The package contents might have shifted, shielding might have been breached, or an error might have occurred in packaging or labeling.

**ACCIDENT-INCIDENT RESPONSE**

It can take days to weeks for the symptoms of a radiological incident to exhibit themselves.  And usually, there won't be any distinctive signatures that such a situation has occurred, since human senses can't detect the colorless and odorless.

But the dimensions of the contaminated area can be determined with specialized equipment. These same devices can also tell whether or not the amounts of radioactivity could possibly result in immediate or long-term health problems. The contaminated area could be larger than that which is apparent. Contaminated people can migrate before the delayed onset of their symptoms has been discovered.

The following discussion is designed to give agencies and individuals key information for responding to a radiological material release.

Remember that your personal safety and the safety of other responders are of the utmost importance as you approach an incident scene to collect data.

But it's equally important to adhere to safety guidelines throughout the entire process of carrying out emergency response tasks.

Another point that can't be overstated: STAY COGNIZANT of the fact that the presence of hazardous radiological materials may not be immediately detected.

YOU MUST KEEP YOUR EXPOSURE LOW!

We all come in contact with some radiation exposure as a natural part of living in our environment. However, it is in your best interest to keep exposure to radiation to a practical minimum.

Here are some basic rules of thumb to follow:

- You'll experience less radiation exposure with the shortest time in a radiation field. One way to keep a person's radiation exposure to a minimum is to have a team that rotates people among tasks.
- Your dose of radiation will be lower with the farthest distance from a radiation source.
- Radioactive substances should NEVER be TOUCHED.
- To escape physical contact, you should use tools, such as brooms, shovels, etc., if you need to move any materials.
- When it's practical within the emergency situation, try to use a barrier to shield yourself and others from the radiation.
- Another way to decrease exposure is to minimize the amount of radioactive substances in the work area.

The CDC (Center for Disease Control and Prevention) has made the following suggestions for responding to a nuclear emergency.
**Find a stable building for shelter. Stay current with national and local emergency-alert information via local television and radio stations.
**Your state and local health agencies will provide some suggestions for protective actions. Follow them.
**Also, adhere to the protective measures (detailed earlier) of minimizing your exposure time to the radiation source, maximizing your distance from the source, and/or remaining behind walls of buildings, or similar physical barriers

What happens if a radiological event involves a nuclear power facility?

- The government will initiate a national emergency response that's been planned and rehearsed for over 20 years.
- Even before such an incident, people living near a nuclear power plant should have been given information regarding that plant's emergency plan.

- Remain inside and avoid opening doors and windows. Keep children indoors.
- Get the ventilation of rooms and buildings under control. Any equipment that brings fresh air in from the outside needs to be turned off, including fans, air conditioners, and forced air heating units. They should only be used to recirculate air that already exists in the facility or structure.
- People outside should get inside the nearest building for immediate shelter.
- If you are going outside for critical or lifesaving purposes, you should cover your mouth and nose. Make sure that you don't stir up and breathe dust particles.
- Keep in mind that you increase your possibility of spreading contamination to others by going outside into the affected area.
- Until the safety of fruits and vegetables grown in the region are deemed to be safe, you should not eat them.

First responders to a nuclear accident must have unique and specific knowledge in order to make preliminary assessments involving this type of disaster.

As for the public and others, personal safety must also be the primary concern for emergency medical personnel as they evaluate and take action against exposure to dangerous levels of radioactivity.

The first responders to the scene of a nuclear weapon or accident scene should set an exclusion zone of 2,500 feet from the source of radiological material.

The only exception to this is when going closer is required to rescue injured people.

No one should try to put out fires or otherwise approach the source of a nuclear incident for any other reason.

In the case of a nuclear weapon, you should immediately report the incident to the nearest military facility and the JNACC (Joint Nuclear Accident Coordinating Center).

Guidelines:
A) It is very unlikely that the radiation material is immediately threatening to the lives of rescuers and victims.
B) Respiratory protection (SC BA) is recommended for emergency response whenever fire, fumes, smoke, or chemicals are involved. It should also be used if environmental conditions could cause material to be airborne.
C) Special protective clothing, turnout gear, or other clothing that covers bare skin can keep contaminants off skin, so they're recommended for emergency response.
D) Contamination with radioactive material does not immediately produce visible skin injury. If redness, blistering, or rash occurs that is not due to heat or flames, you should look for other causes.
E) Exposure does not cause immediate symptoms such as coughing, choking, burning eyes, vomiting, pain, etc., or unconsciousness. Look for other possible causes if you see any of these symptoms.
F) Human exposure can be measured at the accident scene, but make sure that you follow standard protocols, such as those in the U.S. Department of Transportation (DOT) Emergency Response Guidebook.

The following are some indicators that a radiological incident has occurred.
1) Unusual numbers of sick or dying people or animals

First responders should consult with local hospitals to find out if other casualties with similar symptoms have been reported. It may take some hours or days to weeks after the actual incident for casualties to be seen. The radioactive substance used and the dose received will determine the amount of time needed before symptoms become visible. As time progresses, some additional symptoms may appear. Some of these may include skin reddening and vomiting, in severe cases.
2) Metal debris that is out of the ordinary. E.g., materials that appear to be like bombs or other unexplained munitions.
3) Radiation Symbols. Radiation symbols affixed to containers (detailed in a later section).
4) Material that emits heat. Substances that give off heat but don't show any signs of any heat source.
5) Material and/or particles that glow. Strongly radioactive substances can often emit a radio-luminescence.

Response

General protective measures:

- Find a stable building for shelter. Stay current with national and local emergency-alert information via local television and radio stations.
- Your state and local health agencies will provide some suggestions for protective actions. Follow them.
- Adhere to the protective measures of minimizing your exposure time to the radiation source, maximizing your distance from the source, and/or remaining behind walls of buildings, or similar physical barriers. One way to keep a person's radiation exposure to a minimum is to have a team that rotates people among tasks. Radioactive substances should NEVER be TOUCHED. To escape physical contact, you should use tools, such as brooms, shovels, etc., if you need to move any materials. When it's practical within the emergency situation,

try to use a barrier to shield yourself and others from the radiation.

- Remain inside and avoid opening doors and windows. Keep children indoors.
- Get the ventilation of rooms and buildings under control. Any equipment that brings fresh air in from the outside needs to be turned off, including fans, air conditioners, and forced air heating units. They should only be used to recirculate air that already exists in the facility or structure.
- People outside should get inside the nearest building for immediate shelter.
- If going outside for critical or lifesaving purposes, you should cover your mouth and nose. Make sure that you don't stir up and breathe dust particles.
- Until the safety of fruits and vegetables grown in the region are deemed to be safe, you should not eat them.

Emergency Services

The first responders to the scene of a nuclear weapon or accident scene should set an exclusion zone of 2,500 feet from the source of radiological material.

The only exception to this is when going closer is required to rescue injured people.

No one should try to put out fires or otherwise approach the source of a nuclear incident for any other reason.

In the case of a nuclear weapon, you should immediately report the incident to the nearest military facility and the JNACC (Joint Nuclear Accident Coordinating Center). (See the section entitled "Emergency Contacts" for specific phone numbers, etc.).

Respiratory protection is recommended for emergency response whenever fire, fumes, smoke, or chemicals are involved.

Special protective clothing, turnout gear, or other clothing that covers bare skin can keep contaminants off skin, so they're recommended for emergency response.

Contamination with radioactive material does not immediately produce visible skin injury. If redness, blistering, or rash occurs that is not due to heat or flames, you should look for other causes.

Exposure does not cause immediate symptoms such as coughing, choking, burning eyes, vomiting, pain, etc., or unconsciousness. Look for other possible causes if you see any of these symptoms.

Planned procedures should be followed whenever the hospital gets a call to admit a radiation victim.

Whoever receives the call should try to obtain the maximum amount of information:

- How many incident victims there are
- How each victim was injured and their medical status
- Whether or not contamination surveys have been conducted on victims
- The victims' radiological status, that is, whether they've been exposed or contaminated)
- The contaminant's identity, if it can be determined
- The victims' estimated time of arrival to the hospital

Ambulance personnel should be informed of whatever special emergency department entrance exists for radiation incident victims.

Any time that the accident report comes from a source other than usual emergency communications, make sure to get a call-back number.

Then verify the incident before you assemble the radiological emergency response team and prepare for patient admission.

Ambulance personnel should be informed of whatever special emergency department entrance exists for radiation incident victims.

What are some basic emergency care procedures?
A) Create a controlled area that is big enough to hold the number of victims you expect.
B) Put some kind of covering over floors. Keep a watch over control area exits so as to prevent tracking of contaminants.
C) The controlled area should have restrictions that are enforced precisely.
D) Isolation precautions also need to be precisely enforced. Some of these include wearing gloves, masks, boots, gowns, caps, and with double bagging.
E) Establish a buffer zone or secondary control line so as to offer extra security.
F) Control wastes, such as linens, dressings, and clothing, etc., with the use of big, plastic-lined containers.
G) Keep ventilation strictly controlled.
H) Instruments that have become contaminated should be changed. This includes drapes, gloves, etc.
I) Use waterproof materials to restrict the spread of contaminated liquids.

Hospital Strategies

All members of the hospital's emergency response team should know the facility's written plan, and they need to take part in scheduled drills.

Subgroups that handle decontamination, triage, or radiological monitoring and other specialized tasks should conduct quarterly, semiannual or more frequent drills.

*Overview:*

- *Establish a controlled area that's big enough to contain the expected number of patients.*
- *Cover the area floors and monitor controlled area exits to hinder tracking of contaminants.*
- *Minimize controlled area access.*
- *Keep a strict watch on every person and every thing that leaves the controlled area.*
- *Follow isolation precautions strictly, making use of protective clothing and bagging.*
- *To create additional security, use a buffer zone or secondary control line.*
- *Put clothing, linens, and dressings, etc. in large, plastic-lined containers so as to control waste.*
- *Control ventilation.*
- *Change instruments, outer gloves, drapes, etc., when they become contaminated.*
- *Minimize the spread of contaminated liquids by using waterproof aperture drapes and other waterproof materials.*

Radiological emergency response team members need to wear surgical clothing, such as masks, gowns, gloves, eye protection, scrub suits, and caps. This protective clothing can keep contaminants off their bare skin and their own clothing.

Team members also need to wear waterproof shoe covers, and any open seams or cuffs need to be taped shut with adhesive or masking tape. It helps in the tape-removal process if fold-over tabs are placed at the end of each taped area.

Members should wear two pairs of surgical gloves: the first under the arm cuff, secured by tape. The second pair of gloves needs to be worn in a way that makes them easy to take off and replace if they get contaminated.

Every team member needs to be assigned a radiation dosimeter, and it should remain affixed to the outside of the surgical gown. This location makes it easy to read and to remove, if necessary.

Under the surgical gown, members should wear a film badge or other type of dosimeter (if available).

If a team member uses liquids for decontamination purposes, they can also wear a waterproof apron.

Although this protective clothing obstructs alpha and some beta particles, it does not stop gamma rays.

It's best not to use lead aprons like the ones in the x-ray department, since they won't stop most gamma rays either, although they offer a false sense of security to that effect.

How do you prepare the area for contamination control?
A) It's best to choose a treatment room close to an outside entrance, if that is a possibility. Remove all visitors and patients, and clear out or cover every apparatus not needed during the radiation accident patient's emergency care.

B) Get several big waste containers and line them with plastic. Cover the treatment table with several layers of disposable waterproof sheeting. You should have a readily available supply of plastic bags in all sizes.
C) Before the victim is admitted, all survey equipment needs to have been checked and made ready for use. The documentation of background radiation levels should also be completed.
D) The treatment team needs to be ready to join the patient at the ambulance where the victim can be transferred to the prepared treatment gurney.

How do you control the contamination of floor areas?

- You should cover them with rolls of brown butcher paper or wrapping paper, 3 to 4 feet wide. Unrolling them can make a path from the ambulance entrance to the room used for decontamination. (If you can't find paper, it's okay to use regular cloth sheets or square absorbent pads).
- Tape whatever floor covering you use to the floor securely. Then rope off this route and mark it to prohibit the entry of unauthorized personnel. If you have time, cover the decontamination room floor or treatment area floor in the same way. Cleaning up the area will be less complicated this way.
- You should set up a control line at the entrance to the decontamination room. Put a wide strip of tape on the floor at the room's entrance, and mark it so that it's easy to tell the difference between the uncontaminated (non-controlled) side and the contaminated (controlled) side.

Air systems should be shut down.

Prior to the arrival at the hospital, the patient's contaminated clothing needs to be removed (but only if this can happen without resulting in harm or delay.)

If it can't be removed before admission, clothing needs to be taken off as soon as possible after arrival. Extreme caution must be employed to prevent spreading contaminants that may have become embedded in or on the clothing.

A plastic bag should be used to keep all clothing, and any other accompanying blankets, and sheets, etc.

After handling clothing, caregivers need to change gloves or other items that could possibly be contaminated.

Victim Care

**Non-Contaminated Care:**
No specially prepared treatment area is necessary for non-contaminated patients, since they can be given treatment as they would in any other emergency case.

Question the victims as you attend to their medical needs. Find out if you can ascertain whether or not there has been radiation from an external source.

If you know or suspect exposure, order a stat CBC, giving special attention to determining the absolute lymphocyte count.

Make certain that you record the time that the blood sample was extracted.

**Contaminated patient care:**
Keep on attending to the victim's medical needs.

Secure the entire area where victim and emergency personnel have been.

Until cleared by the radiation safety officer, don't allow anyone or anything to be removed from the area.

Set up control lines and stop the spread of contamination.

Determine the victim's radiological status completely.

Make all staff take off contaminated clothing before they leave the area. Next, make them submit to a survey, shower, re-dress in clean clothing, and then submit to another survey before they're allowed to leave the area.
It's possible for contaminated victims to have radioactive materials deposited on skin surfaces or in wounds. They can also have the materials deposited internally, if they've ingested, inhaled, or otherwise absorbed them.

Before giving attention to the patient's radiological status, there should be a re-assessment of the contaminated victim's airway, breathing, and circulation. This should be performed in the decontamination room.

Make a quick determination of the victim's level of consciousness and vital signs, and then stabilize the patient's condition.

Once a full examination of the whole patient has been completed, and all injuries have been identified, a full radiological survey needs to be completed.

Make certain to query the victims about any medications they may be using currently. Also ask whether or not they have any allergies, history of chronic or recent illness, or if they've had any nuclear medicine tests recently.

Be sensitive to the victim's anxiety level and offer psychological support.

Take a full and detailed medical, occupational, and accident history.

Make sure that a comprehensive physical exam has been completed.

---What is the procedure for handling radiological and clinical lab test samples?
MANDATORY: Place all samples in separate, labeled containers that specify name, date, time of sampling, area of samples, and size of area samples.
SUGGESTED: Retain blood, urine, feces, or other samples taken in the emergency treatment period for later examination.
MANDATORY: Get the correct radiation safety, legal, other information so you can properly take care of the requirements for storing and disposing of the samples that have been collected.

Samples Needed:
- CBC and differential STAT (follow with absolute lymphocyte counts every 6 hours for 48 hours when history indicates possibility of total-body irradiation)
- Routine urinalysis
- Swabs from body orifices
- Wound dressing and/or swabs from wounds
- Urine: 24-hour specimen x 4 days
- Feces x 4 days

Guidelines for Admitting:
- Meet the radiation incident patient at either the ambulance or a triage area set up close to the treatment area.

- Make sure that the EMS personnel remain with their vehicle until a radiation safety officer has released them after surveying them, their vehicle, and their equipment.
- If a medical problem is serious, that takes priority over the radiological concerns. Life-threatening problems should receive immediate attention.
- Radiation injury is not immediately life threatening, since it almost never causes unconsciousness or readily apparent signs of injury. For this reason, other causes of injury or illness have to be taken into consideration first.
- If you're not sure about contamination, a critically injured patient needs to be taken immediately into the prepared area.
- If practical in the situation, it's okay to conduct an initial, brief radiological survey to assess contamination. The possibility of contamination exists whenever any radiation survey meter reading measures above background radiation levels.

**Decontamination**

Medical attention may have to be given first to a non-radiological problem if radioactive materials are components of acids, fluorides (uranium hexafluoride-UF6), mercury, lead, or other compounds. This is because some radioactive materials are corrosive or toxic because of their chemical properties.

First, contaminated wounds and body orifices should be decontaminated. Then move on to areas of highest contamination levels on the intact skin.

It's easiest if you wash the contaminated area gently under a stream of water, without splashing.

Simultaneously use a soft brush or surgical sponge to scrub.

Hot water causes vasodilation with increased area blood flow:

- Opens the pores
- Enhances the chance of absorption of the radioactive material through the skin.

Aggressive rubbing tends to cause abrasion and erythema and should be avoided.

Whenever plain water is not effective for washing, you can use a mild soap with a neutral pH. Or you can use surgical scrub soap.

Scrub the area for 3-4 minutes, rinse it for 2-3 minutes, then repeat (if necessary).

Check the contaminated area between each scrub and rinse to determine whether or not there is a decreasing trend with respect to the levels of radiation.

One useful decontamination agent is sodium hypochlorite, diluted 1 to 10 in water.

For calloused areas, you can use a mildly abrasive soap made up of a 1 to 1 mixture of powdered detergent and cornmeal mixed with water into a paste.

Shampoo contaminated hairy parts of the body with several washings.

Clip off contaminated hair if shampooing is unsuccessful.

Don't shave, because even tiny nicks or abrasions can cause internal contamination.

Do everything you can to prevent getting any fluids into the ears, eyes, nose, or mouth when shampooing the head.

++++++++++++

How can ambulatory victims with localized contamination be decontaminated?

- Use a basin or a sink.
- Shower the victim under the scrutiny or with the help of a radiation safety officer, if extensive body areas are contaminated.
- Warn the victim that splashing of water into the eyes, nose, mouth, or ears should be prevented.
- Shower repeatedly, if necessary.
- Dry after each shower with clean towels.

With practicality in mind, decontaminate the patient as thoroughly as possible.

First, limit the spread of radioactivity by draping the wounds (with a waterproof material, if possible).

Gently irrigate the contaminated wound with saline or water to decontaminate it. Repeated irrigations are normally required.

After each irrigation, you should monitor the wound. For accurate results, take off contaminated drapes, dressings, etc., before each monitoring.

Conventional debridement of the wound has to be considered if the contamination level remains seriously high.

Only excise vital tissue upon the expert advise of medical or health physics specialists.

You should retain any debrided or excised tissue for assessment by health physicists.

Remove any visible embedded radioactive materials with forceps or with the use of water-pik.

Use an "en bloc" full thickness biopsy (employing a punch biopsy instrument) to decontaminate puncture wounds containing radioactive particles, especially in the fingers.

What should be done if radioactive material has entered the oral cavity?

- Encourage brushing the teeth with toothpaste and frequent rinsing of the mouth.
- It may be helpful if the victim gargles with a 3-percent hydrogen peroxide solution, if the pharyngeal region is contaminated too.
- If any radioactive substances were swallowed, it may be effective to use a gastric lavage.

How should contaminated eyes be treated?

- Rinse them by directing a stream of water from the inner canthus to the outer canthus of the eye.
- But take care that the nasolcrimal duct doesn't get contaminated in the process.

What should be done about ears that have suffered contamination?

- Contaminated ears require external rinsing, and an ear syringe can be used to rinse the auditory canal, provided the tympanic membrane is intact.

How to do you treat a specific internal contaminant?

| Radionuclide | Medication | For Ingestion/ Inhalation | Principle of Action |
|---|---|---|---|
| Iodine | KI (potassium iodide) | 130 mg (tabl) stat, followed by 130 mg q.d. x 7 if indicated | Blocks thyroid deposition |
| Rare earths<br>Plutonium<br>Transplutonics<br>Yttrium | Zn-DTPA<br>Ca-DTPA | 1 gm Ca-DTPA (Zn-DTPA) in 150-250 ml 5 percent D/W IV over 60 minutes | Chelation |
| Uranium | Bicarbonate | 2 ampules sodium bicarbonate (44.3 mEq each; 7.5%) in 1000 cc normal saline @ 125 cc/hr; alternately, oral | Alkalinization of urine; reduces chance of acute tubular necrosis |

| | | | |
|---|---|---|---|
| | | administration of two bicarbonate tablets every 4 hours until the urine reaches a pH of 8-9 | |
| Cesium<br>Rubidium<br>Thallium | Prussian Blue [Ferrihexacyano-Ferrate (II)] | 1 gm with 100-200 ml water p.o. t.i.d. for several days | Blocks absorption from GI tract and prevents recycling. |
| Tritium | Water | Force fluids | Isotopic dilution |

How does nuclear contamination damage the thyroid gland?

The thyroid gland normally uses iodine from the food we eat, which is non-radioactive. But it can't distinguish between radioactive and non-radioactive forms of iodine.
After a nuclear incident, the thyroid gland will not hesitate to absorb the released radioactive iodine. If that happens, then the thyroid gland's cells can become damaged by radiation from high concentrations of radioactive iodine releases. Thyroid cancer or other diseases of the thyroid can occur within a few years of the exposure, particularly in young children.

What's the procedure for transferring a patient once the emergency treatment is completed?

After decontamination, conduct a final complete-body survey.

- Lay a fresh floor covering from the clean area to the stretcher to be used to move the patient.
- Bring in a clean stretcher and let clean attendants transfer the patient to it. Attendants involved in the decontamination process may be contaminated at this point.
- Wheel the patient to the door.
- Have the radiation officer conduct a final check of both the patient and the stretcher, paying particular attention to the wheels.
- Take the patient from the room.

Before leaving the controlled area, how should decontamination team members remove protective clothes once they've reached the control line?

1. First take off outer gloves, turning them inside out while they're pulled off.
2. Hand the dosimeter to the radiation safety officer.
3. Take off all tape attached to sleeves and trouser cuffs.
4. Take off outer surgical gown, turning it inside-out, making sure not to shake it.
5. Take off surgical pants over shoe covers.
6. Take off mask and head cover.
7. Take off one foot's shoe cover so the radiation safety officer can monitor that shoe.
8. Upon a clean show reading, step across the control line, and allow the radiation safety officer to monitor the second shoe.
9. If both shoes are clean, remove your inner gloves.

Have every team member complete a full-body radiological survey.

Shower thoroughly.

Place a sign outside the door that reads: "CAUTION--CONTROLLED AREA-- DO NOT ENTER"
Keep it secured unless some emergency medical situation requires its use.
Unsecure it only after the radiation safety officer or other health physics expert has checked it and decontaminated it.

<u>Past Catastrophes</u>

Three nuclear incidents will be discussed:

1) Hiroshima (Japan)
2) Nagasaki (Japan)
3) Chernobyl (former Soviet Union)

We will also cover some of what's known about Strontium-90, a radioactive substance that's prevalent in our environment.

Both the Hiroshima and Nagasaki incidents took place in Japan at about the same time. The radiation exposure of both tragedies also resulted in similar effects.

While fighting Japan in World War II, the United States dropped an atomic bomb on Hiroshima on August 6, 1945. Then three days later, the U.S. dropped another one on Nagasaki.

The blasts decimated hundreds of thousands of people in a matter of seconds, along with their houses and places of employment.

Serious injuries occurred, many of which the survivors still suffer from.

As will be outlined in the section on Residual Health Effects," the genetic damages and consequences many never be rectified.

What acute radiation illness resulted from the explosion?
Some of the human damage was from direct thermal injuries, such as flash burns caused by near-ultraviolet rays, visible rays, and infrared rays.

From the least severe to the most severe, flash burns include:
1. Erythema (red burn)
2. Patchy coagulation (spotted white burn)
3. Uniform coagulation (white burn)
4. Steam bleb or blister (blebbed white burn)
5. Carbonization (charred burn)

There were also acute radiation effects, as direct whole-body exposure to ionizing radiation resulted in biological consequences.

From the least dose exposure to the highest dose exposure, some of these included:
* Changes in lymphocytes
* White blood cell loss.
* Nausea and vomiting.
* Suppression of hematopoietic activity (i.e. menstruation disorders).
* Decreased blood cell and platelet formation. Gastrointestinal problems (i.e., diarrhea).
* Coma.
* Death.

The ionizing radiation induced malignant tumors and several cancers among many victims.

Some of them include:

- Salivary gland tumors
- Colon cancer
- Myeloma
- Thyroid cancer
- Breast cancer
- Lung cancer

Damage to the germ cells of the exposed parents' gonads manifested a number of genetic effects. One of the conditions was Down's syndrome, which includes the problems of having an extra copy of chromosome at the 21st position.
In April of 1986, the people at Chernobyl in the Soviet Union experienced the world's worst commercial nuclear accident.

Here, as in the Japan situations, long-range health effects that will persist for generations followed the deaths and injuries.

When the massive concrete lid was inadvertently blown off by one of the four nuclear reactors, a radioactive debris cloud was sent 2000 meters into the air.

A fire started in the buildings surround the core, after the initial blast split the reactor core. For two weeks after, the reactor core burned and radioactive contaminants continued to be set free into the environment.

Cesium-137 accounted for most of the general population's radiation exposure.

Cesium-137 becomes incorporated into the body and begins to influence how tissues and vital organs metabolize. It causes changes to the blood and immune system.

It was incorporated into the food and water ingested by the public, after its easy penetration into the soil and it's absorption by plant life.

Chiefly incorporated by the thyroid, Cesium-137 is also absorbed by the heart, kidneys, liver, brain, spleen, bones, muscles, and other organs.

The accident released 85 P Bq of Cesium-137, but due to the fact that the radiation fallout happened over an extended period of time, it's hard to estimate the exposure dosage for the people in the surrounding distant areas.

What long-term radiation exposure effects from the Chernobyl accident are still being studied today, 15 years after the calamity?

* Cancer within the youth population
* Endocrine deficiencies among children causing obesity or emaciation
* Growth and weight delays
* Mental disorders
* Greater incidences of diseases in the digestive, cardiovascular, nervous, and urinary systems

What long-term blood disorders are caused by radiation contamination?

As a result of the atomic bomb radiation in Japan, a variety of illnesses due to damage to hematopoietic (blood-forming) organs such as the bone marrow and lymph nodes.

Some of them include:

* Infections due to deteriorated defense mechanisms
* Decreased platelets

* Anemia
* Hemorrhaging

What are some indications of the hematopoietic syndrome?

The chief sites of blood cell production are affected: (hematopoiesis)

- Involved are the bone marrow, spleen, and lymph nodes.
- Starts with anorexia or other appetite loss, apathy, nausea, and vomiting after exposure to 2 to 10 grays of radiation.
- Most severe for 6 to 12 hours after contamination, within 24 to 36 hours after exposure these manifestations may completely subside.
- Severe infections occur, even while the symptoms are not displayed, as the victim develops a severe shortage of white blood cells, which fight infections. This is caused as a result of the blood-producing cells in the lymph nodes, spleen, and bone marrow starting to waste away.
- The hematopoietic and gastrointestinal malfunctions are typically fatal with a radiation dose of more than 6.

What can happen with prolonged or repeated exposure to low doses of radiation from radioactive implants or external sources?

* Amenorrhea (a discontinuance of menstrual periods)
* Decreased fertility in men and women
* Diminished libido (sex drive) in women
* Cataracts
* Anemia (diminished number of red blood cells)
* Leukopenia (decreased number of white blood cells
* Thrombocytopenia (Diminished number of platelets)

If extremely large doses of radiation are inflicted upon limited areas of the body, the following symptoms will be seen:

* Loss of hair
* Wasting of skin
* Formation of ulcers (open sores)
* Development of calluses
* Formation of spider veins (small reddened areas of dilated blood vessels just beneath the skin's surface).
* Squamous cell skin cancer can develop over time
* Bone tumors may develop years after the eating of radium salts and some other radioactive compounds.

## Glossary

Acute Radiation Syndrome
Consists of three levels of effects: Hematopoietic (blood cells, most sensitive); Gastrointestinal (GI cells, very sensitive); and Central Nervous System (brain/muscle cells, insensitive). The initial signs and symptoms are nausea, vomiting, fatigue, and loss of appetite. Below about 200 rems, these symptoms may be the only indication of radiation exposure.

alpha particle (a)
A specific particle ejected from a radioactive atom. It has low penetrating power and short range. Alpha particles will generally fail to penetrate the skin. Alpha-emitting atoms can cause health effects if introduced into the lungs or wounds.

The alpha particle has a very short range in air and a very low ability to penetrate other materials, but it has a strong ability to ionize materials. Alpha particles are unable to penetrate even the thin layer of dead cells of human skin and consequently are not an external radiation hazard. Alpha-emitting nuclides inside the body as a result of inhalation or ingestion are a considerable internal radiation hazard.

atom
The smallest piece of an element that cannot be divided or broken up by chemical means.

background radiation
The radiation in man's natural environment, including cosmic rays and radiation from the naturally radioactive elements, both outside and inside the bodies of humans and animals. It is also called natural radiation. Man-made sources of radioactivity contribute to total background radiation levels.

Becquerel
The SI unit of activity 1 disintegration per second; 37 billion Bq = 1 curie. (See conversion factors in the Measurement <http://www.orau.gov/reacts/measure.htm> section.)

beta particles (b)
A small particle ejected from a radioactive atom. It has a moderate penetrating power and a range of up to a few meters in air. Beta particles will penetrate only a fraction of an inch of skin tissue. High-energy electrons emitted from the nucleus of an atom during radioactive decay. They normally can be stopped by the skin or a very thin sheet of metal.

Cesium-137 (Cs-137)
A strong gamma ray source and can contaminate property, entailing extensive clean-up. It is commonly used in industrial measurement gauges and for irradiation of material. Half-life is 30.2 years.

Cobalt-60 (Co-60)

A strong gamma ray source, and is extensively used as a radiotherapeutic for treating cancer, food and material irradiation, gamma radiography, and industrial measurement gauges. Half-life is 5.27 years.

contamination
Deposition of radioactive material in any place where it is not desired, particularly where its presence can be harmful.

controlled area
An area where entry, activities, and exit are controlled to help ensure radiation protection and prevent the spread of contamination.

cosmic rays
High-energy radiation that originates outside the Earth's atmosphere.

curie (Ci)
A unit of measure used to describe the amount of radioactivity in a sample of material. A unit of radioactive decay rate defined as $3.7 \times 10^{10}$ disintegrations per second.

decay
The process by which an unstable element is changed to another isotope or another element by the spontaneous emission of radiation from its nucleus. This process can be measured by using radiation detectors such as Geiger counters.

decontamination
The reduction or removal of contaminating radioactive material from a structure, area, object, or person. The process of making people, objects, or areas safe by absorbing, destroying,

neutralizing, making harmless, or removing the hazardous material.

detector
A device that is sensitive to radiation and can produce a response signal suitable or measurement or analysis. A radiation detection instrument.

dose
A general term for the amount of radiation or energy absorbed over a period of time.

dose rate
The dose delivered per unit of time. It is usually expressed as rads per hour or in multiples or submultiples of this unit such as millirads per hour. The dose rate is commonly used to indicate the level of hazard from a radioactive source.

dosimeter
A small, pocket-sized device used for monitoring radiation exposure of personnel. Before use, it is given a charge, and the amount of discharge that occurs is a measure of the accumulated radiation exposure.
A portable instrument for measuring and registering the total accumulated dose to ionizing radiation.

electromagnetic radiation
A traveling wave motion that results from changing electric and magnetic fields. Types of electromagnetic radiation range from those of short wavelength, like x rays and gamma rays, through the ultraviolet, visible, and infrared regions, to radar and radio waves of relatively long wavelengths.

exposure

A quantity used to indicate the amount of ionization in air produced by x- or gamma-ray radiation. The unit is the roentgen (R). For practical purposes, one roentgen is comparable to 1 rad or 1 rem for X and gamma radiation. The SI unit of exposure is the coulomb per kilogram (C/kg). One R = 2.58 x $10^{-4}$ C/kg.

gamma rays, or gamma radiation (g)
High-energy photons emitted from the nucleus of atoms; similar to x rays. They can penetrate deeply into body tissue and many materials. Cobalt-60 and Cesium-137 are both strong g-emitters. Shielding against gamma radiation requires thick layers of dense materials, such as lead. Gamma rays are potentially lethal to humans.

Geiger counter or G-M meter
An instrument used to detect and measure radiation.

Gray
The SI unit of absorbed dose; 1 gray = 100 rads.

half-life
The amount of time needed for half of the atoms of a radioactive material to decay.

Highly Enriched Uranium (HEU)
Uranium that is enriched to above 20% Uranium-235 (U-235). Weapons-grade HEU is enriched to above 90% in U-235.

Inverse square law
The relationship that states that electromagnetic radiation intensity is inversely proportional to the square of the distance from a point source.

ionize
To split off one or more electrons from an atom, thus leaving it with a positive electric charge. The electrons usually attach to one of the atoms or molecules, giving them a negative charge.

ionization
Production of charged particles in a medium. An orbital electron is stripped from a neutral atom, producing an ion pair (a negatively charged electron and a positively charged atom).

ionizing radiation
Electromagnetic (X ray and gamma) or particulate (alpha, beta) radiation capable of producing ions or charged particles.

Iridium-192
A gamma-ray emitting radioisotope used for gamma-radiography. The half-life is 73, 83 days.

irradiation
Exposure to ionizing radiation.

isotope
A specific element always has the same number of protons in the nucleus. That same element may, however, appear in forms that have different numbers of neutrons in the nucleus. These different forms are referred to as "isotopes" of the element. For example, deuterium (2H) and tritium (3H) are isotopes of ordinary hydrogen (H).

lethal dose (50/30)
The dose of radiation expected to cause death within 30 days to 50% of those exposed without medical treatment. The generally accepted range from 400-500 rem received over a short period of time.

monitoring
Determining the amount of ionizing radiation or radioactive contamination present. Also referred to as surveying.

nuclear reactor
A device in which a controlled, self-sustaining nuclear chain reaction can be maintained with the use of cooling to remove generated heat.

Plutonium-239 (Pu-239)
A metallic element used for nuclear weapons. The half-life is 24,110 years.

Rad
The unit of radiation absorbed dose.
A unit of absorbed dose of radiation defined as deposition of 100 ergs of energy per gram of tissue. It amounts to approximately one ionization per cubic micron.

radiation
Energy traveling through space. Some types of radiation associated with radioactivity are alpha and beta particles and gamma and X rays.
High energy alpha or beta particles or gamma rays that are emitted by an atom as the substance undergoes radioactive decay.

radiation sickness
Symptoms resulting from excessive exposure to radiation of the body.

radioactive waste

Disposable, radioactive materials resulting from nuclear operations. Wastes are generally classified into two categories, high-level and low-level waste.

radioactivity
The spontaneous emission of radiation from the nucleus of an unstable atom. As a result of this emission, the radioactive atom is converted, or decays, into an atom of a different element that might or might not be radioactive.

Radiological Dispersal Device (RDD)
A device (weapon or equipment), other than a nuclear explosive device, designed to disseminate radioactive material in order to cause destruction, damage, or injury by means of the radiation produced by the decay of such material.

radioluminescence
The luminescence produced by particles emitted during radioactive decay.

rem
A measure of radiation dose related to biological effect.
A Roentgen Man Equivalent is a unit of absorbed dose that takes into account the relative effectiveness of radiation that harms human health.

Roentgen
The unit of exposure from X or gamma rays (see exposure).

sealed source
A radioactive source, sealed in an impervious container that has sufficient mechanical strength to prevent contact with and dispersion of the radioactive material under the conditions of use and wear for which it was designed. Generally used for

radiography or radiation therapy. May be classified "Special Form" on shipping papers and packages.

shielding
Materials (lead, concrete, etc.) used to block or attenuate radiation for protection of equipment, materials, or people.

Sievert
The SI unit of dose equivalent; 1 Sv = 100 rem.

Special Nuclear Material (SNM)
Plutonium and uranium enriched in the isotope Uranium-233 or Uranium 235.

Uranium 235 (U-235)
Naturally occurring uranium U-235 is found at 0.72% enrichment. U-235 is used as an reactor fuel or for weapons; however, weapons typically use U-235 enriched to 90%. The half-life is 7.04 x $10^8$ years.

X-Ray
Penetrating electromagnetic radiation whose wavelengths are shorter than those of visible light. An invisible, highly penetrating electromagnetic radiation of much shorter wavelength (higher frequency) than visible light. Very similar to gamma-rays.

## OSHA Regulations on Ionizing Radiation

1910.1096(a)

Definitions applicable to this section.

1910.1096(a)(1)

Radiation includes alpha rays, beta rays, gamma rays, X-rays, neutrons, high-speed electrons, high-speed protons, and other atomic particles; but such term does not include sound or radio waves, or visible light, or infrared or ultraviolet light.

1910.1096(a)(2)

Radioactive material means any material which emits, by spontaneous nuclear disintegration, corpuscular or electromagnetic emanations.

1910.1096(a)(3)

Restricted area means any area access to which is controlled by the employer for purposes of protection of individuals from exposure to radiation or radioactive materials.

1910.1096(a)(4)

Unrestricted area means any area access to which is not controlled by the employer for purposes of protection of individuals from exposure to radiation or radioactive materials.

..1910.1096(a)(5)

1910.1096(a)(5)

Dose means the quantity of ionizing radiation absorbed, per unit of mass, by the body or by any portion of the body. When the provisions in this section specify a dose during a period of time, the dose is the total quantity of radiation absorbed, per unit of mass, by the body or by any portion of the body during such period of time. Several different units of dose are in current use. Definitions of units used in this section are set forth in paragraphs (a)(6) and (7) of this section.

1910.1096(a)(6)

Rad means a measure of the dose of any ionizing radiation to body tissues in terms of the energy absorbed per unit of mass of the tissue. One rad is the dose corresponding to the absorption of 100 ergs per gram of tissue (1 millirad (mrad)=0.001 rad).

1910.1096(a)(7)

Rem means a measure of the dose of any ionizing radiation to body tissue in terms of its estimated biological effect relative to a dose of 1 roentgen (r) of X-rays (1 millirem (mrem)=0.001 rem). The relation of the rem to other dose units depends upon the biological effect under consideration and upon the conditions for irradiation. Each of the following is considered to be equivalent to a dose of 1 rem:

1910.1096(a)(7)(i)

A dose of 1 roentgen due to X- or gamma radiation;

1910.1096(a)(7)(ii)

A dose of 1 rad due to X-, gamma, or beta radiation;

1910.1096(a)(7)(iii)

A dose of 0.1 rad due to neutrons or high energy protons;

1910.1096(a)(7)(iv)

A dose of 0.05 rad due to particles heavier than protons and with sufficient energy to reach the lens of the eye;

..1910.1096(a)(7)(v)

1910.1096(a)(7)(v)

If it is more convenient to measure the neutron flux, or equivalent, than to determine the neutron dose in rads, as provided in paragraph (a)(7)(iii) of this section, 1 rem of neutron radiation may, for purposes of the provisions in this section be assumed to be equivalent to 14 million neutrons per square centimeter incident upon the body; or, if there is sufficient information to estimate with reasonable accuracy the approximate distribution in energy of the neutrons, the incident number of neutrons per square centimeter equivalent to 1 rem may be estimated from Table G-17:

TABLE G-17 - NEUTRON FLUX DOSE EQUIVALENTS

| Neutron energy (million electron volts (Mev)) | Number of neutrons per square centimeter equivalent to a dose of 1 rem (neutrons/cm(2)) | Average flux to deliver 100 millirem in 40 hours (neutrons/cm(2) per sec) |
|---|---|---|
| Thermal | 970 X 10(6) | 670 |
| 0.0001 | 720 X 10(6) | 500 |
| 0.005 | 820 X 10(6) | 570 |
| 0.02 | 400 X 10(6) | 280 |
| 0.1 | 120 X 10(6) | 80 |
| 0.5 | 43 X 10(6) | 30 |
| 1.0 | 26 X 10(6) | 18 |
| 2.5 | 29 X 10(6) | 20 |
| 5.0 | 26 X 10(6) | 18 |

| | | |
|---|---|---|
| 7.5 | 24 X 10(6) | 17 |
| 10 | 24 X 10(6) | 17 |
| 10 to 30 | 14 X 10(6) | 10 |

1910.1096(a)(8)

For determining exposures to X- or gamma rays up to 3 Mev., the dose limits specified in this section may be assumed to be equivalent to the "air dose". For the purpose of this section air dose means that the dose is measured by a properly calibrated appropriate instrument in air at or near the body surface in the region of the highest dosage rate.

1910.1096(b)

Exposure of individuals to radiation in restricted areas.

1910.1096(b)(1)

Except as provided in paragraph (b)(2) of this section, no employer shall possess, use, or transfer sources of ionizing radiation in such a manner as to cause any individual in a restricted area to receive in any period of one calendar quarter from sources in the employer's possession or control a dose in excess of the limits specified in Table G-18:

TABLE G-18

| | Rems per calendar quarter |
|---|---|
| Whole body: Head and trunk; active blood-forming organs; lens of eyes; or gonads. | 1 1/4 |
| Hands and forearms; feet and ankles | 18 3/4 |
| Skin of whole body | 7 1/2 |

1910.1096(b)(2)

An employer may permit an individual in a restricted area to receive doses to the whole body greater than those permitted under subparagraph (1) of this paragraph, so long as:

1910.1096(b)(2)(i)

During any calendar quarter the dose to the whole body shall not exceed 3 rems; and

..1910.1096(b)(2)(ii)

1910.1096(b)(2)(ii)

The dose to the whole body, when added to the accumulated occupational dose to the whole body, shall not exceed 5 (N-18) rems, where "N" equals the individual's age in years at his last birthday; and

1910.1096(b)(2)(iii)

The employer maintains adequate past and current exposure records which show that the addition of such a dose will not cause the individual to exceed the amount authorized in this subparagraph. As used in this subparagraph Dose to the whole body shall be deemed to include any dose to the whole body, gonad, active bloodforming organs, head and trunk, or lens of the eye.

1910.1096(b)(3)

No employer shall permit any employee who is under 18 years of age to receive in any period of one calendar quarter a dose in excess of 10 percent of the limits specified in Table Z-18.

1910.1096(b)(4)

Calendar quarter means any 3-month period determined as follows:

1910.1096(b)(4)(i)

The first period of any year may begin on any date in January: Provided, That the second, third, and fourth periods accordingly begin on the same date in April, July, and October, respectively, and that the fourth period extends into January of the succeeding year, if necessary to complete a 3-month quarter. During the first year of use of this method of determination, the first period for that year shall also include any additional days in January preceding the starting date for the first period; or

..1910.1096(b)(4)(ii)

1910.1096(b)(4)(ii)

The first period in a calendar year of 13 complete, consecutive calendar weeks; the second period in a calendar year of 13 complete, consecutive weeks; the third period in a calendar year of 13 complete, consecutive calendar weeks; the fourth period in a calendar year of 13 complete, consecutive calendar weeks. If at the end of a calendar year there are any days not falling within a complete calendar week of that year, such days shall be included within the last complete calendar week of that year. If at the beginning of any calendar year there are days not falling within a complete calendar week of that year, such days shall be included within the last complete calendar week of the previous year; or

1910.1096(b)(4)(iii)

The four periods in a calendar year may consist of the first 14 complete, consecutive calendar weeks; the next 12 complete, consecutive calendar weeks, the next 14 complete, consecutive calendar weeks, and the last 12 complete, consecutive calendar weeks. If at the end of a calendar year there are any days not falling within a complete calendar week of that year, such days shall be included (for purposes of this section) within the last complete calendar week of the year. If at the beginning of any calendar year there are days not falling within a complete calendar week of that year, such days shall be included (for purposes of this section) within the last complete week of the previous year.

1910.1096(c)

Exposure to airborne radioactive material.

..1910.1096(c)(1)

1910.1096(c)(1)

No employer shall possess, use or transport radioactive material in such a manner as to cause any employee, within a restricted area, to be exposed to airborne radioactive material in an average concentration in excess of the limits specified in Table 1 of appendix B to 10 CFR part 20. The limits given in Table 1 are for exposure to the concentrations specified for 40 hours in any workweek of 7 consecutive days. In any such period where the number of hours of exposure is less than 40, the limits specified in the table may be increased proportionately. In any such period where the number of hours of exposure is greater than 40, the limits specified in the table shall be decreased proportionately.

1910.1096(c)(2)

No employer shall possess, use, or transfer radioactive material in such a manner as to cause any individual within a restricted area, who is under 18 years of age, to be exposed to airborne radioactive material in an average concentration in excess of the limits specified in Table II of appendix B to 10 CFR part 20. For purposes of this paragraph, concentrations may be averaged over periods not greater than 1 week.

1910.1096(c)(3)

Exposed as used in this paragraph means that the individual is present in an airborne concentration. No allowance shall be made for the use of protective clothing or equipment, or particle size.

1910.1096(d)

Precautionary procedures and personal monitoring.

1910.1096(d)(1)

Every employer shall make such surveys as may be necessary for him to comply with the provisions in this section. Survey means an evaluation of the radiation hazards incident to the production, use, release, disposal, or presence of radioactive materials or other sources of radiation under a specific set of conditions. When appropriate, such evaluation includes a physical survey of the location of materials and equipment, and measurements of levels of radiation or concentrations of radioactive material present.

..1910.1096(d)(2)

1910.1096(d)(2)

Every employer shall supply appropriate personnel monitoring equipment, such as film badges, pocket chambers, pocket dosimeters, or film rings, and shall require the use of such equipment by:

1910.1096(d)(2)(i)

Each employee who enters a restricted area under such circumstances that he receives, or is likely to receive, a dose in any calendar quarter in excess of 25 percent of the applicable value specified in paragraph (b)(1) of this section; and

1910.1096(d)(2)(ii)

Each employee under 18 years of age who enters a restricted area under such circumstances that he receives, or is likely to receive, a dose in any calendar quarter in excess of 5 percent of the applicable value specified in paragraph (b)(1) of this section; and

1910.1096(d)(2)(iii)

Each employee who enters a high radiation area.

1910.1096(d)(3)

As used in this section:

1910.1096(d)(3)(i)

Personnel monitoring equipment means devices designed to be worn or carried by an individual for the purpose of measuring the dose received (e.g., film badges, pocket chambers, pocket dosimeters, film rings, etc.);

1910.1096(d)(3)(ii)

Radiation area means any area, accessible to personnel, in which there exists radiation at such levels that a major portion of the body could receive in any 1 hour a dose in excess of 5 millirem, or in any 5 consecutive days a dose in excess of 100 millirem; and

..1910.1096(d)(3)(iii)

1910.1096(d)(3)(iii)

High radiation area means any area, accessible to personnel, in which there exists radiation at such levels that a major portion of the body could receive in any one hour a dose in excess of 100 millirem.

1910.1096(e)

Caution signs, labels, and signals -

1910.1096(e)(1)

General.

1910.1096(e)(1)(i)

Symbols prescribed by this paragraph shall use the conventional radiation caution colors (magenta or purple on yellow background). The symbol prescribed by this paragraph is the conventional three-bladed design:

FIGURE G-10 RADIATION SYMBOL
(For Figure G-10, Click Here)

1910.1096(e)(2)

Radiation area. Each radiation area shall be conspicuously posted with a sign or signs bearing the radiation caution symbol described in subparagraph (1) of this paragraph and the words:

CAUTION

RADIATION AREA

1910.1096(e)(3)

High radiation area.

1910.1096(e)(3)(i)

Each high radiation area shall be conspicuously posted with a sign or signs bearing the radiation caution symbol and the words:

CAUTION

HIGH RADIATION AREA

..1910.1096(e)(3)(ii)

1910.1096(e)(3)(ii)

Each high radiation area shall be equipped with a control device which shall either cause the level of radiation to be reduced below that at which an individual might receive a dose of 100 millirems in 1 hour upon entry into the area or shall energize a conspicuous visible or audible alarm signal in such a manner that the individual entering and the employer or a supervisor of the activity are made aware of the entry. In the case of a high radiation area established for a period of 30 days or less, such control device is not required.

1910.1096(e)(4)

Airborne radioactivity area.

1910.1096(e)(4)(i)

As used in the provisions of this section, airborne radioactivity area means:

1910.1096(e)(4)(i)(a)

Any room, enclosure, or operating area in which airborne radioactive materials, composed wholly or partly of radioactive material, exist in concentrations in excess of the amounts specified in column 1 of Table 1 of appendix B to 10 CFR part 20 or

1910.1096(e)(4)(i)(b)

Any room, enclosure, or operating area in which airborne radioactive materials exist in concentrations which, averaged over the number of hours in any week during which individuals are in the area, exceed 25 percent of the amounts specified in column 1 of Table 1 of appendix B to 10 CFR part 20.

1910.1096(e)(4)(ii)

Each airborne radioactivity area shall be conspicuously posted with a sign or signs bearing the radiation caution symbol described in paragraph (e)(1) of this section and the words:

CAUTION

AIRBORNE RADIOACTIVITY AREA

1910.1096(e)(5)

Additional requirements.

..1910.1096(e)(5)(i)

1910.1096(e)(5)(i)

Each area or room in which radioactive material is used or stored and which contains any radioactive material (other than natural uranium or thorium) in any amount exceeding 10 times the quantity of such material specified in appendix C to 10 CFR part 20 shall be conspicuously posted with a sign or signs bearing the radiation caution symbol described in paragraph (e)(1) of this section and the words:

CAUTION

RADIOACTIVE MATERIALS

1910.1096(e)(5)(ii)

Each area or room in which natural uranium or thorium is used or stored in an amount exceeding 100 times the quantity of such material specified in 10 CFR part 20 shall be conspicuously posted with a sign or signs bearing the radiation caution symbol described in paragraph (e)(1) of this section and the words:

CAUTION

RADIOACTIVE MATERIALS

1910.1096(e)(6)

Containers.

1910.1096(e)(6)(i)

Each container in which is transported, stored, or used a quantity of any radioactive material (other than natural uranium or thorium) greater than the quantity of such material specified in appendix C to 10 CFR part 20 shall bear a durable, clearly visible label bearing the radiation caution symbol described in paragraph (e)(1) of this section and the words:

CAUTION

RADIOACTIVE MATERIALS

1910.1096(e)(6)(ii)

Each container in which natural uranium or thorium is transported, stored, or used in a quantity greater than 10 times the quantity specified in appendix C to 10 CFR part 20 shall bear a durable, clearly visible label bearing the radiation caution symbol described in paragraph (e)(1) of this section and the words:

CAUTION

RADIOACTIVE MATERIALS

1910.1096(e)(6)(iii)

Notwithstanding the provisions of paragraphs (e)(6)(i) and (ii) of this section a label shall not be required:

..1910.1096(e)(6)(iii)(a)

1910.1096(e)(6)(iii)(a)

If the concentration of the material in the container does not exceed that specified in column 2 of Table 1 of appendix B to 10 CFR part 20, or

1910.1096(e)(6)(iii)(b)

For laboratory containers, such as beakers, flasks, and test tubes, used transiently in laboratory procedures, when the user is present.

1910.1096(e)(6)(iv)

Where containers are used for storage, the labels required in this subparagraph shall state also the quantities and kinds of radioactive materials in the containers and the date of measurement of the quantities.

1910.1096(f)

Immediate evacuation warning signal -

1910.1096(f)(1)

Signal characteristics.

1910.1096(f)(1)(i)

The signal shall be a midfrequency complex sound wave amplitude modulated at a subsonic frequency. The complex sound wave in free space shall have a fundamental frequency (f(1)) between 450 and 500 hertz (Hz) modulated at a subsonic rate between 4 and 5 hertz.

1910.1096(f)(1)(ii)

The signal generator shall not be less than 75 decibels at every location where an individual may be present whose immediate, rapid, and complete evacuation is essential.

..1910.1096(f)(1)(iii)

1910.1096(f)(1)(iii)

A sufficient number of signal units shall be installed such that the requirements of paragraph (f)(1)(ii) of this section are met at every location where an individual may be present whose immediate, rapid, and complete evacuation is essential.

1910.1096(f)(1)(iv)

The signal shall be unique in the plant or facility in which it is installed.

1910.1096(f)(1)(v)

The minimum duration of the signal shall be sufficient to insure that all affected persons hear the signal.

1910.1096(f)(1)(vi)

The signal-generating system shall respond automatically to an initiating event without requiring any human action to sound the signal.

1910.1096(f)(2)

Design objectives.

1910.1096(f)(2)(i)

The signal-generating system shall be designed to incorporate components which enable the system to produce the desired signal each time it is activated within one-half second of activation

1910.1096(f)(2)(ii)

The signal-generating system shall be provided with an automatically activated secondary power supply which is adequate to simultaneously power all emergency equipment to which it is connected, if operation during power failure is necessary, except in those systems using batteries as the primary source of power.

..1910.1096(f)(2)(iii)

1910.1096(f)(2)(iii)

All components of the signal-generating system shall be located to provide maximum practicable protection against damage in case of fire, explosion, corrosive atmosphere, or other environmental extremes consistent with adequate system performance.

1910.1096(f)(2)(iv)

The signal-generating system shall be designed with the minimum number of components necessary to make it function as intended, and should utilize components which do not require frequent servicing such as lubrication or cleaning.

1910.1096(f)(2)(v)

Where several activating devices feed activating information to a central signal generator, failure of any activating device shall not render the signal-generator system inoperable to activating information from the remaining devices.

1910.1096(f)(2)(vi)

The signal-generating system shall be designed to enhance the probability that alarm occurs only when immediate evacuation is warranted. The number of false alarms shall not be so great that the signal will come to be disregarded and shall be low enough to minimize personal injuries or excessive property damage that might result from such evacuation.

1910.1096(f)(3)

Testing.

1910.1096(f)(3)(i)

Initial tests, inspections, and checks of the signal-generating system shall be made to verify that the fabrication and installation were made in accordance with design plans and specifications and to develop a thorough knowledge of the performance of the system and all components under normal and hostile conditions.

..1910.1096(f)(3)(ii)

1910.1096(f)(3)(ii)

Once the system has been placed in service, periodic tests, inspections, and checks shall be made to minimize the possibility of malfunction.

1910.1096(f)(3)(iii)

Following significant alterations or revisions to the system, tests and checks similar to the initial installation tests shall be made.

1910.1096(f)(3)(iv)

Tests shall be designed to minimize hazards while conducting the tests.

1910.1096(f)(3)(v)

Prior to normal operation the signal-generating system shall be checked physically and functionally to assure reliability and to demonstrate accuracy and performance. Specific tests shall include:

1910.1096(f)(3)(v)(a)

All power sources.

1910.1096(f)(3)(v)(b)

Calibration and calibration stability.

1910.1096(f)(3)(v)(c)

Trip levels and stability.

1910.1096(f)(3)(v)(d)

Continuity of function with loss and return of required services such as AC or DC power, air pressure, etc.

1910.1096(f)(3)(v)(e)

All indicators.

1910.1096(f)(3)(v)(f)

Trouble indicator circuits and signals, where used.

..1910.1096(f)(3)(v)(g)

1910.1096(f)(3)(v)(g)

Air pressure (if used)

1910.1096(f)(3)(v)(h)

Determine that sound level of the signal is within the limit of paragraph (f)(1)(ii) of this section at all points that require immediate evacuation.

1910.1096(f)(3)(vi)

In addition to the initial startup and operating tests, periodic scheduled performance tests and status checks must be made to insure that the system is at all times operating within design limits and capable of the required response. Specific periodic tests or checks or both shall include:

1910.1096(f)(3)(vi)(a)

Adequacy of signal activation device.

1910.1096(f)(3)(vi)(b)

All power sources.

1910.1096(f)(3)(vi)(c)

Function of all alarm circuits and trouble indicator circuits including trip levels.

1910.1096(f)(3)(vi)(d)

Air pressure (if used).

1910.1096(f)(3)(vi)(e)

Function of entire system including operation without power where required.

..1910.1096(f)(3)(vi)(f)

1910.1096(f)(3)(vi)(f)

Complete operational tests including sounding of the signal and determination that sound levels are adequate.

1910.1096(f)(3)(vii)

Periodic tests shall be scheduled on the basis of need, experience, difficulty, and disruption of operations. The entire system should be operationally tested at least quarterly.

1910.1096(f)(3)(viii)

All employees whose work may necessitate their presence in an area covered by the signal shall be made familiar with the actual sound of the signal-preferably as it sounds at their work location. Before placing the system into operation, all employees normally working in the area shall be made acquainted with the signal by actual demonstration at their work locations.

1910.1096(g)

Exceptions from posting requirements. Notwithstanding the provisions of paragraph (e) of this section:

1910.1096(g)(1)

A room or area is not required to be posted with a caution sign because of the presence of a sealed source, provided the radiation level 12 inches from the surface of the source container or housing does not exceed 5 millirem per hour.

..1910.1096(g)(2)

1910.1096(g)(2)

Rooms or other areas in onsite medical facilities are not required to be posted with caution signs because of the presence of patients containing radioactive material, provided that there are personnel in attendance who shall take the precautions necessary to prevent the exposure of any individual to radiation or radioactive material in excess of the limits established in the provisions of this section.

1910.1096(g)(3)

Caution signs are not required to be posted at areas or rooms containing radioactive materials for periods of less than 8 hours: Provided, That

1910.1096(g)(3)(i)

The materials are constantly attended during such periods by an individual who shall take the precautions necessary to prevent the exposure of any individual to radiation or radioactive materials in excess of the limits established in the provisions of this section; and

1910.1096(g)(3)(ii)

Such area or room is subject to the employer's control.

1910.1096(h)

Exemptions for radioactive materials packaged for shipment. Radioactive materials packaged and labeled in accordance with regulations of the Department of Transportation published in 49 CFR Chapter I, are exempt from the labeling and posting requirements of this subpart during shipment, provided that the inside containers are labeled in accordance with the provisions of paragraph (e) of this section.

..1910.1096(i)

1910.1096(i)

"Instruction of personnel, posting."

1910.1096(i)(1)

Employers regulated by the Nuclear Regulatory Commission shall be governed by 10 CFR part 20 standards. Employers in a State named in paragraph (p)(3) of this section shall be governed by the requirements of the laws and regulations of that State. All other employers shall be regulated by the following:

1910.1096(i)(2)

All individuals working in or frequenting any portion of a radiation area shall be informed of the occurrence of radioactive materials or of radiation in such portions of the radiation area; shall be instructed in the safety problems associated with exposure to such materials or radiation and in precautions or devices to minimize exposure; shall be instructed in the applicable provisions of this section for the protection of employees from exposure to radiation or radioactive materials; and shall be advised of reports of radiation exposure which employees may request pursuant to the regulations in this section.

1910.1096(i)(3)

Each employer to whom this section applies shall post a current copy of its provisions and a copy of the operating procedures applicable to the work conspicuously in such locations as to insure that employees working in or frequenting radiation areas will observe these documents on the way to and from their place of employment, or shall keep such documents available for examination of employees upon request.

1910.1096(j)

Storage of radioactive materials. Radioactive materials stored in a non-radiation area shall be secured against unauthorized removal from the place of storage.

..1910.1096(k)

1910.1096(k)

Waste disposal. No employer shall dispose of radioactive material except by transfer to an authorized recipient, or in a manner approved by the Nuclear Regulatory Commission or a State named in paragraph (p)(3) of this section.

1910.1096(l)

Notification of incidents -

1910.1096(l)(1)

Immediate notification. Each employer shall immediately notify the Assistant Secretary of Labor or his duly authorized representative, for employees not protected by the

Nuclear Regulatory Commission by means of 10 CFR part 20; paragraph (p)(2) of this section, or the requirements of the laws and regulations of States named in paragraph (p)(3) of this section, by telephone or telegraph of any incident involving radiation which may have caused or threatens to cause:

1910.1096(l)(1)(i)

Exposure of the whole body of any individual to 25 rems or more of radiation; exposure of the skin of the whole body of any individual to 150 rems or more of radiation; or exposure of the feet, ankles, hands, or forearms of any individual to 375 rems or more of radiation; or

1910.1096(l)(1)(ii)

The release of radioactive material in concentrations which, if averaged over a period of 24 hours, would exceed 5,000 times the limit specified for such materials in Table II of appendix B to 10 CFR part 20.

..1910.1096(l)(2)

1910.1096(l)(2)

Twenty-four hour notification. Each employer shall within 24 hours following its occurrence notify the Assistant Secretary of Labor or his duly authorized representative for employees not protected by the Nuclear Regulatory Commission by means of 10 CFR part 20; paragraph (p)(2) of this section, or the requirements of the laws and applicable regulations of States named in paragraph (p)(3) of this section, by telephone or telegraph of any incident involving radiation which may have caused or threatens to cause:

1910.1096(l)(2)(i)

Exposure of the whole body of any individual to 5 rems or more of radiation; exposure of the skin of the whole body of any individual to 30 rems or more of radiation; or exposure of the feet, ankles, hands, or forearms to 75 rems or more of radiation; or

1910.1096(l)(2)(ii)

[Reserved]

1910.1096(m)

"Reports of overexposure and excessive levels and concentrations."

1910.1096(m)(1)

In addition to any notification required by paragraph (1) of this section each employer shall make a report in writing within 30 days to the Assistant Secretary of Labor or his duly authorized representative, for employees not protected by the Nuclear Regulatory

Commission by means of 10 CFR part 20; or under paragraph (p)(2) of this section, or the requirements of the laws and regulations of States named in paragraph (p)(3) of this section, of each exposure of an individual to radiation or concentrations of radioactive material in excess of any applicable limit in this section. Each report required under this paragraph shall describe the extent of exposure of persons to radiation or to radioactive material; levels of radiation and concentration of radioactive material involved, the cause of the exposure, levels of concentrations; and corrective steps taken or planned to assure against a recurrence.

..1910.1096(m)(2)

1910.1096(m)(2)

In any case where an employer is required pursuant to the provisions of this paragraph to report to the U.S. Department of Labor any exposure of an individual to radiation or to concentrations of radioactive material, the employer shall also notify such individual of the nature and extent of exposure. Such notice shall be in writing and shall contain the following statement: "You should preserve this report for future reference."

1910.1096(n)

"Records."

1910.1096(n)(1)

Every employer shall maintain records of the radiation exposure of all employees for whom personnel monitoring is required under paragraph (d) of this section and advise each of his employees of his individual exposure on at least an annual basis.

1910.1096(n)(2)

Every employer shall maintain records in the same units used in tables in paragraph (b) of this section and appendix B to 10 CFR part 20.

1910.1096(o)

"Disclosure to former employee of individual employee's record."

1910.1096(o)(1)

At the request of a former employee an employer shall furnish to the employee a report of the employee's exposure to radiation as shown in records maintained by the employer pursuant to paragraph (n)(1) of this section. Such report shall be furnished within 30 days from the time the request is made, and shall cover each calendar quarter of the individual's employment involving exposure to radiation or such lesser period as may be requested by the employee. The report shall also include the results of any calculations and analysis of radioactive material deposited in the body of the employee. The report shall be in writing and contain the following statement: "You should preserve this report for future reference."

..1910.1096(p)

1910.1096(p)

"Nuclear Regulatory Commission licensees - NRC contractors operating NRC plants and facilities - NRC Agreement State licensees or registrants."

1910.1096(p)(1)

Any employer who possesses or uses source material, byproduct material, or special nuclear material, as defined in the Atomic Energy Act of 1954, as amended, under a license issued by the Nuclear Regulatory Commission and in accordance with the requirements of 10 CFR part 20 shall be deemed to be in compliance with the requirements of this section with respect to such possession and use.

1910.1096(p)(2)

NRC contractors operating NRC plants and facilities: Any employer who possesses or uses source material, byproduct material, special nuclear material, or other radiation sources under a contract with the Nuclear Regulatory Commission for the operation of NRC plants and facilities and in accordance with the standards, procedures, and other requirements for radiation protection established by the Commission for such contract pursuant to the Atomic Energy Act of 1954 as amended (42 U.S.C. 2011 et seq.), shall be deemed to be in compliance with the requirements of this section with respect to such possession and use.

1910.1096(p)(3)

NRC-agreement State licensees or registrants:

..1910.1096(p)(3)(i)

1910.1096(p)(3)(i)

"Atomic Energy Act sources." Any employer who possesses or uses source material, byproduct material, or special nuclear material, as defined in the Atomic Energy Act of 1954, as amended (42 U.S.C. 2011 et seq.), and has either registered such sources with, or is operating under a license issued by, a State which has an agreement in effect with the Nuclear Regulatory Commission pursuant to section 274(b) (42 U.S.C. 2021(b)) of the Atomic Energy Act of 1954, as amended, and in accordance with the requirements of that State's laws and regulations shall be deemed to be in compliance with the radiation requirements of this section, insofar as his possession and use of such material is concerned, unless the Secretary of Labor, after conference with the Nuclear Regulatory Commission, shall determine that the State's program for control of these radiation sources is incompatible with the requirements of this section. Such agreements currently are in effect only in the States of Alabama, Arkansas, California, Kansas, Kentucky, Florida, Mississippi, New Hampshire, New York, North Carolina, Texas, Tennessee, Oregon, Idaho, Arizona, Colorado, Louisiana, Nebraska, Washington, Maryland, North Dakota, South Carolina, and Georgia.

1910.1096(p)(3)(ii)

"Other sources." Any employer who possesses or uses radiation sources other than source material, byproduct material, or special nuclear material, as defined in the Atomic Energy Act of 1954, as amended (42 U.S.C. 2011 et seq.), and has either registered such sources with, or is operating under a license issued by a State which has an agreement in effect with the Nuclear Regulatory Commission pursuant to section 274(b) (42 U.S.C. 2021(b)) of the Atomic Energy Act of 1954, as amended, and in accordance with the requirements of that State's laws and regulations shall be deemed to be in compliance with the radiation requirements of this section, insofar as his possession and use of such material is concerned, provided the State's program for control of these radiation sources is the subject of a currently effective determination by the Assistant Secretary of Labor that such program is compatible with the requirements of this section. Such determinations currently are in effect only in the States of Alabama, Arkansas, California, Kansas, Kentucky, Florida, Mississippi, New Hampshire, New York, North Carolina, Texas, Tennessee, Oregon, Idaho, Arizona, Colorado, Louisiana, Nebraska, Washington, Maryland, North Dakota, South Carolina, and Georgia.

[39 FR 23502, June 27, 1974, as amended at 43 FR 49746, Oct. 24, 1978; 43 FR 51759, Nov. 7, 1978; 49 FR 18295, Apr. 30, 1984; 58 FR 35309, June 30, 1993; 61 FR 5507, Feb. 13, 1996; 61 FR 31427, June 20, 1996]

OSHA Regulations on Non-Ionizing Radiation

1910.97(a)

"Electromagnetic radiation" -

1910.97(a)(1)

"Definitions applicable to this paragraph."

1910.97(a)(1)(i)

The term "electromagnetic radiation" is restricted to that portion of the spectrum commonly defined as the radio frequency region, which for the purpose of this specification shall include the microwave frequency region.

1910.97(a)(1)(ii)

"Partial body irradiation." Pertains to the case in which part of the body is exposed to the incident electromagnetic energy.

1910.97(a)(1)(iii)

"Radiation protection guide." Radiation level which should not be exceeded without careful consideration of the reasons for doing so.

1910.97(a)(1)(iv)

The word "symbol" as used in this specification refers to the overall design, shape, and coloring of the rf radiation sign shown in figure G-11.

..1910.97(a)(1)(v)

1910.97(a)(1)(v)

"Whole body irradiation." Pertains to the case in which the entire body is exposed to the incident electromagnetic energy or in which the cross section of the body is smaller than the cross section of the incident radiation beam.

1910.97(a)(2)

"Radiation protection guide."

1910.97(a)(2)(i)

For normal environmental conditions and for incident electromagnetic energy of frequencies from 10 MHz to 100 GHz, the radiation protection guide is 10 mW/cm.(2) (milliwatt per square centimeter) as averaged over any possible 0.1-hour period. This means the following:

Power density: 10 mW./cm.(2) for periods of 0.1-hour or more.
Energy density: 1 mW.-hr./cm.(2) (milliwatt hour per square centimeter) during any 0.1-hour period.

This guide applies whether the radiation is continuous or intermittent.

1910.97(a)(2)(ii)

These formulated recommendations pertain to both whole body irradiation and partial body irradiation. Partial body irradiation must be included since it has been shown that some parts of the human body (e.g., eyes, testicles) may be harmed if exposed to incident radiation levels significantly in excess of the recommended levels.

1910.97(a)(3)

"Warning symbol."

1910.97(a)(3)(i)

The warning symbol for radio frequency radiation hazards shall consist of a red isosceles triangle above an inverted black isosceles triangle, separated and outlined by an aluminum color border. The words "Warning - Radio-Frequency Radiation Hazard" shall appear in the upper triangle. See figure G-11.

..1910.97(a)(3)(ii)

1910.97(a)(3)(ii)

American National Standard Safety Color Code for Marking Physical Hazards and the Identification of Certain Equipment, Z53.1-1953 which is incorporated by reference as specified in Sec. 1910.6, shall be used for color specification. All lettering and the border shall be of aluminum color.

1910.97(a)(3)(iii)

The inclusion and choice of warning information or precautionary instructions is at the discretion of the user. If such information is included it shall appear in the lower triangle of the warning symbol.

FIGURE G-11
RADIO-FREQUENCY RADIATION HAZARD WARNING SYMBOL
(For Figure G-11, Click Here)

1910.97(a)(4)

"Scope." This section applies to all radiations originating from radio stations, radar equipment, and other possible sources of electromagnetic radiation such as used for communication, radio navigation, and industrial and scientific purposes. This section does not apply to the deliberate exposure of patients by, or under the direction of, practitioners of the healing arts.

[61 FR 9227, March 7, 1996]